£29.95

Practical Ideas

for Metalworking Operations,
Tooling, and Maintenance

Practical Ideas

for Metalworking Operations, Tooling, and Maintenance

A compendium of shop methods, recipes, tooling innovations, and manufacturing techniques contributed by the readers of AMERICAN MACHINIST as an aid to solving problems and to improving productivity.

By the editors of **AMERICAN MACHINIST**

American Machinist
McGraw-Hill Publications Company
New York

Library of Congress Cataloging in Publication Data

Main entry under title:

Practical ideas—for metalworking operations, tooling, and maintenance.

"A compendium of shop methods, recipes, tooling innovations, and manufacturing techniques contributed by the readers of American Machinist as an aid to solving problems and to improving productivity."
 Includes index.
 1. Metal-work. 2. Production engineering. I. American Machinist (New York, N.Y.: 1968)
TS213.P67   1984   621.8 84-24497
ISBN 0-07-001551-1 (McGraw-Hill)

Copyright © 1984 by McGraw-Hill Inc. All rights reserved. Printed in the United States of America. No part of this publication may be reproduced, stored in a retrieval system, or transmitted in any form or by any means; electronic, mechanical, photocopying, recording, or otherwise, without the prior written permission of the publisher.
1234567890

The sponsoring editor of this book is William M. Stocker, Jr.

Cover design by Arthur Finklestein.

Table of Contents

    Preface vi
    Index ix
1 Drilling 1
2 Milling and Boring 27
3 Turning 69
4 Grinding and Finishing 135
5 Cutoff and Sawing 155
6 Threading 163
7 Tools and Tooling 181
8 Forming and Press Tooling 197
9 Bench Work 245
10 Layout, Inspection, and Measurement 255
11 Joining and Assembly 293
12 Maintenance and Repair 307
13 Miscellaneous 313

PREFACE

Nearly 100 years ago *American Machinist* introduced a new department. It was titled "Letters from Practical Men" and it focused attention upon the wealth of practical shop experiences contributed by AM's readers. From the very first issue of *American Machinist* in 1877 readers' correspondence has provided an ongoing opportunity for them to share their wealth of skills and experiences with other members of this huge manufacturing society we call "metalworking." About 40 years ago, what had become "Ideas from Practical Men" was renamed "Practical Ideas" and it has consistently won the acclaim of AM's readers.

America's reputation for manufacturing know-how for so many decades has resulted from a combination of many ingredients. Not the least of these donations over the years has been the generosity with which so many skilled people have shared their knowledge and their ideas. By training and educating others, especially the young people who aspire to these careers, they have helped to build a nation of problem solvers.

Another strength is seen in the steady flow of talented people who have emigrated to the United States from the world's nations, bringing with them countless special skills. That contribution continues today. Further strength stems from the powerful drive among both the new and the native-born Americans to improve both their own and their families' social and economic positions as well as to better the opportunities for their descendants. And the simple satisfaction of solving a problem or overcoming an obstacle cannot be ignored.

If there is one great force which contributes to so-called American ingenuity, it is the openness with which people, regardless of position, can contribute their ideas to meeting challenges. Their willingness to give of their experience and innovative ideas in solving industry's problems is encouraged by a simple right—freedom of self-expression. Add to this source, nevertheless, one more asset. It has been said that nowhere in the world have the companies that employ these people been as generous over the years in allowing their employees to share this wealth as in the United States. This exchange of information strengthens the whole industrial structure as well as the entire national economy. Ultimately, the total effect is beneficial. It is an old adage that, "One lets more light in by raising the window blind than is let out."

Preface

Good ideas are everywhere, in every endeavor. Some of the contributors to this book live in other parts of the world. These pages include a relatively special range of "Practical Ideas... for Operations, Tooling, and Maintenance" in the metalworking industries, primarily, as well as in repair shops and other places where metals are worked. The book includes ideas submitted by readers and published in *American Machinist* during the last five years. Descriptions, usually illustrated with sketches or drawings, vary from simple shop "kinks" to fairly elaborate manufacturing procedures and equations.

The value of one idea, however, is not in its sophistication but in its ability to solve an immediate, or a recurring, problem. Ideas thrive where minds are open. Often, simply searching for a solution or glancing at another method will trigger an analogy or a spontaneous answer that is different from anything in print. A detailed Index together with the classifying of operations should simplify the search. And we hope that you will find the pages of *American Machinist* a continuing source of guidance.

"Practical Ideas..." was compiled as an aid to everyone who is responsible for working metals—or metal-related materials—as well as for students of these skills and techniques. America's metalworking industries employ some 10-million people, most of them in manufacturing operations. This book offers a selection of ideas which can help the majority of these people as well as millions more in other nations. Within these pages are more than 500 specific solutions to shop problems. From these ideas, new answers may arise. Problem solving never stops.

The editors of American Machinist dedicate this book to the thousands of contributors who, over the years, have helped so many people by sharing their know-how with AM and its readers. In the words of an editor published many years ago, we would describe these contributors and their readers as, the *people who make dreams come true*. We also especially thank AM's former Senior Editor Robert L. Hatschek who was responsible for AM's "Practical Ideas" department during the period when the ideas in this book first appeared. Despite Bob's recent retirement, he is continuing to edit these contributions as a special assignment.

Our appreciation also goes to Publication Services, of Urbana, Illinois, the company that designed, organized and produced this book; especially to Lori Martinsek, who coordinated these efforts.

Today we see the skills of yesteryear taking new directions. Whether craftspeople become "programmers" or continue as manual artists, the need for fresh, practical ideas will always endure. One purpose of this book is to provide a place to start.

WILLIAM M. STOCKER, JR., DIRECTOR
AMERICAN MACHINIST INFORMATION SERVICES

INDEX

Accordian die, 8.02
Adaptable die, 8.28
Adapter plate
   for jig borer, 2.01
   to extend table tooling area, 2.02
Allen wrenches, 7.06
Angle, adjustable, 10.01
Angle-block, adjustable, for milling vise, 2.03
Angled operations
   grinding on surface grinder, 4.12, 4.15
   protractor used for, 1.18
   setup for, 2.47, 2.63
Angle dresser, for surface grinder, 4.01
Angle iron, made into angle plate, 7.22
Angle plates
   made from angle iron, 7.22
   used to make V-block, 10.02
   versatile, 10.45
Arbor press
   lathe used as, 11.04
   punching in, 8.42
   used to cut keyways, 7.04
Arbor
   expanding to grip workpiece, 3.34
   for fragile workpieces, 4.19
   removing from drill chuck, 1.36
   self-locking, 3.68
   split expansion, 3.82
   split spacer for 2.54
   sprag, 3.83
   stepped, 8.58
   tapered, 1.36
Arc, large radius, cutting, 2.14
Automatic toolchanger, 2.58, 10.10

Backfacing jobs, oversize, 2.52
Ball centers, for grinding tapers, 4.03
Ball-turning attachment, turret-lathe, 3.09
Band iron, for hole layout, 1.04
Bandsaw blades
   carbide guides for, 5.02
   welding, 5.14
Bandsaw welder, to anneal spring ends, 13.03

Bar-puller attachment, turret-lathe, 3.10, 3.11
Barstock soft jaws, 3.12
Bearing installation, 11.21
Bearings, replacement in blind holes, 12.05
Bench, adjustable edge, 9.03
Bench blocks, non-tipping, 7.14
Blades, straightening warped, 13.44
Blind dowel pins, 9.18
Blind holes
   clearing chips from, 9.07
   pulling dowels from, 11.27
   removing broken drill from, 13.30
   replacement of bearings in, 12.05
Blueprint storage, 13.50
Bolt circles
   diameter of odd-numbered, 10.60
   microcomputer program for, 10.29
Bolt-hole indexing, 1.30
Bore gage, extension for, 10.04
Bore size, 3.72
Boring bar support, on lathe, 3.14, 3.76
Boring machines, cleaning floor-plate T-slots, 2.40
Boring, soft jaw, 3.69
Boring/threading bar, 6.02
Boring tools
   holder for, 2.48
   setting diameter of, 2.30
Broaching slots, on NC machine, 2.50
Bushing bar, to drill overlapping holes, 1.40
Bushing plug, to prevent press distortion, 11.01
Bushings
   aligning, 11.04
   collet, 3.22
   installing difficult to reach, 11.09
   used as dowel, 7.17
   used for eccentric turning, 3.17
   V-bottom, to center rough castings, 1.41

Calculators, used for layout, 10.23
Caliper readings, 10.28

Capscrews
  eccentric, workpieces clamped with, 7.25
  hard-to-reach, 8.01
Carbide guides, for bandsaw blades, 5.02
Carbide inserts, bench fixture for, 7.01
Carbide scraper, 9.10
C-clamps, 7.13
Center drill
  converted to radius tool, 3.63
  holder, on lathe toolblock, 3.20
Center-drilling
  both ends of rods simultaneously, 1.05
  toolholder to control depth in, 3.95
Center drill tip, removing from workpiece, 13.06
Center-finder, plastic, 10.35
Centerline scriber, spotting punch as, 10.52
Center location, transfer of, 10.58
Chamfer and radius, 10.05
Chamfer, hole, 7.16
Channel, converted into stepped strap clamps, 7.22
Chasers, recycled, 6.15
Chip-disposal problems, 5.06
Chuck
  drill, 1.07, 1.08, 1.36, 1.39, 13.11
  flywheel to spin down, 3.38
  holding lathe center in, 3.06
  magnetic, 4.16, 4.21
  parts catcher for, 3.56
  platform, 1.24
  protection of, 3.19, 13.27
Chucking
  four jaw, 3.40, 3.57, 3.74
  of thin disks, 3.21, 3.66
  magnetic spacers used for, 3.53
  taps, 6.03
Chuck jaws
  brass face for, 3.15
  grinding, 3.41
Circle scriber, 10.08
Clamping, heavy, 2.60
Clamps
  "C," 7.13
  for nonmagnetic parts, 4.04
  for odd shaped workpieces, 2.08
  hand, 9.08
  hose, to hold delicate work, 2.26
  parallel, 9.13, 9.19
  quill, 2.55

  self-adjusting, 7.03
  strap, 7.22, 7.23
  Super Glue used as, 13.18
  use on magnetic-base electric drillpress, 7.02
Collet bushing, to hold stubby parts, 3.22
Collet pads, held with soft jaws, 3.78
Collets
  5C, bench holder for, 3.13
  for toolholding, 3.23
  replacement for, 13.48
Collet stop, adjustable, 3.04
Color coding, 13.08
  for similar components, 11.03
Combination square, 10.09
Compressive stress, in diesets, 8.24
Computer. *See* Microcomputer
Concave radius, use of rotary table, 3.67
Concentricity, proximity probe to find, 10.36
Coolant
  air as, 3.24
  and drilling tools, 1.22
  system, 13.09
Copy-milling cams, on vertical mill, 2.12
Counterbore
  drill inserted in, 1.11
  modified, as countersink, 1.29
Countersink
  counterbore modified as, 1.29
  depth, figuring, 10.11
  diameters, gage to measure, 10.16
Crossbar handles, of drilling chuck keys, knobs for, 13.46
Cross-drilling
  of screws, 1.23
  threaded parts, 1.26
  turned parts, 3.26
Cross-slide, calibrated, 3.18
Crowned work, milling with flycutter, 2.34
Crowns, grinding, 4.11
Cutoff, part-holder for, 3.55
Cutting fluid can, magnet for, 13.20
Cutting fluids, aerator to preserve, 13.34
Cylinders, hollow, support for, 3.90
Cylinders, small, sizing of, 11.13
Cylindrical grinder, used with surface grinder, 4.23

Deburring
  holes, 9.06

Index

keyways, 9.04
of thin materials, 1.10
slots, 9.04
station, 13.01
tool, rotary, 9.12
while machining, 2.16
Deburring machines, media tubs for, 4.05
Deep-hole drill, 1.13
Dental burrs, as milling cutters, 2.17
Depth stops, on drillpresses, 1.06
Dial caliper
   resetting, 12.03
   to set combination square, 10.09
Dial indicator
   holder for, 10.03
   to check automatic toolchanger, 10.10
Diamond grinding wheel, 4.02, 4.14
Diehead, floating, 6.14
Die, holder, 6.10, 6.11, 8.14
Die inserts, press brake tooling, 8.15
Dies
   accordian, 8.02
   adaptable, 8.28
   adjustable, 8.04
   air-bending, 8.12
   bend angle of, 8.26
   clearance of, 8.54
   combination, 8.10
   double-action blanking-and-cupping, 8.17
   draw, 8.25
   embossing, 8.22
   for flat rolling, 8.47
   maintenance of, 8.08
   notching, 8.41
   offset, adjustable, 8.16
   piercing, 8.37
   pillars, realigning, 8.44
   redesign of, 8.40
   roughened, to control metal flow, 8.46
   second-operation, for accurate parts, 8.49
   single, 8.55
   swagging, 8.03
   to produce 10-in-dia holes, 8.35
Die-set
   casters, 5.09
   compressive stress in, 8.24
   jammed, 8.06
Diestock spinner, 6.06
Dowel pins, blind, 9.18
Drawbar wrench, as soft hammer, 2.18

Draw die, 8.25
Drawings, long, held with brackets, 13.05
Draw-punch, wear of, 8.33
Dressing tool, 4.18
Drill
   attachments, to improve accuracy, 1.03
   chatter problems with, 1.34
   deep-hole, 1.13
   depth, marked by magnet, 3.52
   inserted in counterbore, 1.11
   manual, power feed for, 1.01
   point depth, setting, 10.43
   set with planer gage, 2.45
   size, finding, 10.14
   spade, chatter problems, 1.31
   straight-flute, 4.20
   twist, polishing flutes of, 1.32
Drill chucks
   closing jaws of, 1.07
   modified for safe removal, 13.11
   positioning method for shaft centering, 1.08
   removing tapered arbors from, 1.36
Drill drift, adding slide hammer to, 1.39
Drilling. *See also* Center-drilling;
   Cross-drilling
   from tailstock, 3.07
   of large drums, 1.37
   sheetmetal, 1.38
Drilling tools, and coolant, 1.22
Drill jig
   fitted to machine vise, 1.15
   locators for, 1.27
Drillpress
   depth stops on, 1.06
   hand-tapping stations for, 6.09
   jacking up, 1.25
   mini, 1.28
   mounted on vertical mill, 1.21
   recessing tool for, 1.35
   slow-speed, 1.09
   spindle, 1.33
   tables, raising of, 1.12
   used to punch thin stock, 1.14
Drill rack, 1.19

Eccentric capscrews, workpieces clamped with, 7.25
Eccentric reliefs, cut on form cutter, 3.48
Eccentric spade drill, for ID-thread, 3.31

Eccentric turning, 3.17, 3.71, 7.08
  in three-jaw check, 3.32, 3.61
Edge finders, 10.57
Edge stop, for vertical milling machine, 3.33
EDM operation, guide plates for, 7.10
End mill
  corner-rounding, as lathe tool, 3.25
  cutting holes larger with, 7.21
  indicator adapter attachment, 10.21
Eyebolt, to fit three different threads, 13.12

Face cam, 3.35
File cleaning, 13.39
Fixture holder, 3.36
Fixtures, drilling, air-bearing base for, 1.02
Flat belts, made from strapping tape, 13.13
Flycutter
  adjustable, 1.16
  for crowned work, 2.34
  three-bit, 2.61
Flycutting
  of odd slots, 2.22
  Woodruff-style keyways, 7.09
Flywheel
  to speed tailstock, 3.37
  to spin down chuck, 3.38
Foamed plastic. *See* Plastic, foamed
Form cutter, eccentric reliefs cut on, 3.48
Forming, spring plungers align part for, 8.56
Form-tools
  adjustable, 3.02
  surface grinder used for, 4.21
Four-jaw chuck, 3.40, 3.57, 3.74
  plug for, 3.57
Friction pin, 11.05

Gang-milling, backup vise jaw for, 2.7
Gaskets, cutting, 5.04
Grease gun, mini, from small syringe, 13.38
Grinding draft clearance, 13.29
Grinding, multipart fixture for, 7.18
Grinding wheel
  diamond, 4.02, 4.14
  dressing, 4.07
Groove diameter, internal, 10.5, 10.07
Grooving tool, 3.29, 3.77

Hacksaw blade, conversion to pull saw, 9.11
Hacksawing, of round steel barstock, 5.13

Hammer, soft
  drawbar wrench as, 2.18
  preserving, 13.26
Hand clamps, 9.08
Hand-press, double-acting, 8.18
Hand screwmachine
  cross-drilling turned parts in, 3.26
  stock advance for, 3.86
Hand stamping, backward, 9.02
Hand-tapping stations, added to drillpress, 6.09
Heat-shrink part protectors, 13.14
Heat sink, for grinding small punches, 4.13
Heat-treating, of thin-blades, 13.28
Heat-warped parts, 13.24
Height setter, lathe tool, 3.50
Hex-driver, 11.06, 11.11
Hex-key/pin-punch combination, 8.29
Hex, milling from lathe turret, 2.33
Hinges, double rolled, 8.27
Hobbing, spline on overlong shaft, 7.11
Hole chamfer, used for locating and clamping, 7.16
Hole layout, band iron for, 1.04
Hole-punching tool. 8.49
Hole-saw
  chips, disposal of, 5.06
  improving stability of, 5.10
Holes, cutting larger with end mill, 7.21
Horizontal mill, to turn spherical parts, 2.24
Horseshoe milling spacers, 2.25
Hose clamps
  retaining rings as, 13.31
  to hold delicate work, 2.26
ID grooving, with Woodruff cutter, 3.44
ID-thread, eccentric spade drill for, 3.31
Indicator adapter
  for end mills, 10.21
  for vertical mills, 2.28
Indicator mount, magnetic, 10.18
Internal collet, 3.89
Internal groove diameter, 10.05, 10.07

Jackshaft, suspended, 12.06
Jaws
  changing on large vise, 9.01
  for long jobs, 3.85
  of milling vises, 2.37
Jig borer, adapter plate for, 2.01

Index

Jig, to bandsaw short keys, 5.08

Kennedy toolbox, adding drawer to, 13.2
Keys, short, bandsawing 5.08
Keyways
  cut on lathe, 2.13
  cutting oversize, 2.15
  cut with arbor press, 7.04
  deburring, 9.04
  depth, gage to measure, 10.17
  offset, gage to measure, 10.46
  spur gear used to mill, 2.56
  Woodruff-style, 7.09

Lapping, with toothpaste, 13.47
Lathe
  accurate depths on, 3.60
  boring-bar support on, 3.14, 3.76
  dog, for miniature work, 3.59
  dog, to grip narrow flanges, 3.49
  keyways, cut on, 2.13
  long-bed, rebuilding, 3.62
  rim-rolling job on, 8.32
  shortening screws on, 13.33
  tool positioning on, 3.70
  used as horizontal arbor press, 11.14
  vertical mill used as, 3.79
  worn, rebores own tailstock, 12.08
Lathe toolblock, centerdrill holder on, 3.20
Lathe tool, corner-rounding end-mill as, 3.25
Lathe turret
  milling a hex from, 2.33
  Morse-taper toolholders for, 3.54
Layout
  calculators used for, 10.23
  measuring machine used for, 10.27
Length gage, 10.24
Light bulb, used in soldering leaks, 11.15
Long-pitch helixes, milling, 6.05

Machine dials, calibrating, 10.48
Machine jack, miniature, 7.12
Machine table, masking tape on to reduce plate noise, 13.22
Machine vise, drill jig fitted to, 1.15
Machining
  angled, protractor used for, 1.18
  deburring while, 2.16
Machining center, plug for, 2.42

Magnetic chucks, 4.21
  backstop for, 4.16
  wear of, 4.26
Magnetic spacers, chucking, 3.53
Magnet, to mark drill depth, 3.52
Mandrel, 3.34, 3.89
  threaded, removing workpieces from, 3.65
Marking pressure, estimating, 8.30
Masking tape, on machine table, to reduce noise of plates, 13.22
Mating spur gears, for drilling in tight corners, 1.17
Metal stampings, made with single-action press, 8.19
Microcomputer program, for bolt circles, 10.29
Micrometer fixture, 10.30, 10.31
Mill, horizontal, to turn spherical parts, 2.24
Milling cutters, dental burrs as, 2.17
Milling machine crank, mitten for, 13.10
Milling-machine shanks, tool-stand for, 2.62
Milling-machine table
  adapter plate to extend area of, 2.02
  extended for splining long shaft, 2.32
  multipurpose bars for, 2.39
  precision fences for, 2.43
  square setups on, 2.36
Milling plates square, 2.35
Milling radii, without rotary table, 2.21
Milling spacers, horseshoe, 2.25
Milling vise
  adding attachments to, 2.06
  adjustable angle block for, 2.03
  expanding capacity for flat work, 2.20
  handle for, 2.27
  jaws of, 2.37
  mounting off center, 2.38
  setup for angled operations, 2.47
Mirror, magnetic, 13.21
Morse-taper toolholders, 3.54

NC cutters, 2.23
NC cycle time, 3.96
NC machine
  broaching slots on, 2.50
  subplate on, 2.59
NC mill, conversion to NC flamecutter, 5.03
NC punching tool, adaptations of, 8.34
Notching die, 8.41

Offset dies, adjustable, 8.16
Offset drivers, for screws and nuts, 11.07
O-rings, 6.14, 12.04
Overhead weld, at floor level, 11.18
Oversize workpieces, grinding, 4.24

Parallel clamps, 7.23, 9.13, 9.19
Parallelogram case, used for centering, 10.33
Parallelogram T-nuts, 2.40
Parallels
   adjustable, 10.34
   holding in vise, 13.15
Part protectors, heat-shrink, 13.14
Photographs
   of setups, 13.17
   used in printed-circuit-board assembly, 11.10
Piercing die, 8.37
Pin-presser, 11.08
Pipe stops, adjustable, to eliminate workpiece deflection, 2.04
Pipe-turning spider, 3.08, 3.81
Pistons, for heavy clamping, 2.60
Pitches, long, roll feed for, 8.43
Planer gage, set drill with, 2.45
Plastic belting, used as polishing tool, 9.15
Plastic center-finder, 10.35
Plastic, foamed
   cutting, 5.05
   fabricating, 5.01
   machining, 3.51
   turning, 3.94
Plastic strapping, used as shims, 13.25
Plastic tubing, for positioning small nuts and screws, 11.19
Platform chuck, 1.24
Pliers, modified for pulling, 11.20
Precision fences, for milling table, 2.43
Press brake tooling, 8.15
Printed-circuit-board assembly, 11.10
Projection welding, of aluminum, 11.24
Protractor
   reversible, 10.39
   used for angled machining, 1.18
Punches
   bevel extends life of, 8.05
   broken, 8.07, 8.13
   fine, surface grinding, 4.22
   forming, 8.48
   grinding perforating, 4.06
   heat-sink for grinding, 4.13
   spotting, 10.52
   steel stamping, 9.16
   undersize, 8.63
Punching
   from inside to ease notching, 8.41
   in arbor press, 8.42
   second operation, 8.51
   see-through stripper to simplify, 8.52
   thin stock, with drillpress, 1.14
   tool, adaptation of, 8.34
Punch mark, centering under spindle, 2.11
Punch pad, plastic, 8.39

Quill clamp, 2.55
Quill-locking level, holding loose, 2.09
Quill-stop, for milling machines, 2.41
Quotations, preparing, 13.19

Radiator hoses, sawing, 5.11
Radii, milling without rotary table, 2.21
Radius and chamfer, equivalent table for, 10.05
Radiused workpiece, grinding, 4.08
Radius, finding, of a segment, 10.37
Radius tool, converted from center drill, 3.63
Radius-turning tools, holder for, 3.80
Recessing tool, for drillpress, 1.35
Retaining-ring grooves, in shafts, 3.29
Retaining rings, as hose-clamps, 13.31
Rim-rolling job, on lathe, 8.32
Rocking fixture for working positioning, 1.37
Roller tool, used to straighten bent shaft, 7.15
Rolling tool, used to cut stainless steel, 2.44
Root diameter, gage to measure, 10.49
Rotary table, fitting with center plug, 2.10

Scale, reconditioning of, 10.38
Scraper, carbide, 9.10
Screw diameter, 10.12
Screw jack. tubular, adjustable, 10.56
Screws
   cross-drilling of, 1.23
   grinding, 7.20
   locking with rubber plugs, 11.26
   method for locking, 11.17
   shortening on lathe, 13.33
Scriber, 10.41, 10.52

Index

and step-block, give built-in dimensions, 10.54
for straight or curved edge, 10.40
Sealant, for welds, 11.23
Setscrews, as transfer punches, 10.53
Setup jack, miniature, 7.12
Shaft centering, 1.08
Shafts
  polishing, 3.46
  straightening, 7.15
  stub ends, 3.57
Shanks
  flattening, 4.09
  tool tray for, 2.19
Sheetmetal bend, 13.40
Sheetmetal, drilling, 1.38
Shimming, on four dimensions, 2.57
Shim, thin, 13.49
Sine bar, made with dowel pins, 10.47
Sizing, small cylinders, 11.13
Sleeves, fragile puller for, 11.22
Slide-hammer, adding to drill drift, 1.39
Slip spindle plates, 1.40
Slitting saw
  as grooving tool, 3.77
  as lathe parting-tool blade, 3.16
Slots
  broaching on NC machine, 2.50
  deburring, 9.04
Slugs, 8.31
Snap gage, for obstructions, 10.50
Soft jaws
  barstock, 3.12
  boring, 3.69
  from wood, 13.51
  hold collet pads, 3.78
  used to grip turning fixtures, 3.30
Spacer rods, adjustable, 9.05
Spacers
  for depth adjustment, 2.51
  for straddle milling, 2.05
  split, for milling arbor, 2.54
  used with micrometer or fixed stops, 3.05
Spade drill
  and chatter problems, 1.31
  restoring, 4.17
Spherical parts, horizontal mill to turn, 2.24
Spider
  pipe-turning, 3.08, 3.81
  two-legged, 3.98

Spindle, centering punch mark under, 2.11
Spinner, diestock, 6.06
Spline, hobbing, on overlong shaft, 7.11
Splining long shaft, mill table extended for, 2.32
Split spacers, for milling arbor, 2.54
Split stamping, 8.38
Spool, as guide, 13.42
Spotting punch, as centerline scriber, 10.52
Sprag arbor, 3.83
Spray bottle, used for lubricants, 13.16
Spring ends, annealing, with bandsaw welder, 13.03
Spring plungers, align part for forming, 8.56
Springs
  preventing slipping of, 11.12
  winding, 3.100
Spring wire, straighten in lathe, 13.35
Spur gear, to mill keyways, 2.56
Square, adjustable, additional angles for, 10.32
Square boxes, trimming, 8.62
Square holes, machining with round cutters, 2.31
Squareness, indicator stand to check, 10.22
Square setups, on milling table, 2.36
Square stock, holding in three-jaw chuck, 3.43
Stainless steel plugs, ending pullup with, 8.23
Stamping
  hand, 9.02
  reinforcing, 8.45
Steady-rest setup, gage for, 3.73
Step-block shims, 2.57
Stepdrill, automatic toolchanger capacity, 2.58
Stepped arbor, 8.58
Stock guide
  cammed, 8.09
  to oil work, 8.59
Stock, heavy, 8.57
Stop
  feeds narrower strip, 8.21
  for both end operation, 7.19
  trigger, 8.64
Stop-block, and scriber, give built-in dimensions, 10.54
Straddle milling, adjustable spacer for, 2.05
Straight-flute drills, 4.20
Straightness, gage to check, 10.15

Strap clamps
  converted from channel, 7.22
  used to make parallel clamps, 7.23
Strip feed accuracy, 8.20
Stripper, see-through, 8.52
Stud, broken, removal of, 13.7
Styrene test, for NC tape-proofing, 8.60
Styrofoam. *See* Plastic, foamed
Subplate, on NC machine, 2.59
Super Glue, used as clamp, 13.18
Surface grinder
  angle dresser for, 4.01
  grinding fine punches, 4.22
  getting stock square in, 4.10
  grinding compound angles on, 4.12, 4.15
  grinding form-tools with, 4.21
  used with cylindrical grinder, 4.23
  vertical spindle for, 4.25
Swagging dies, 8.03
Swivel vise, setting up, 2.46

Table cleanup, 13.41
Tab stop, 8.61
Tailstock center, disk on, 3.27
Tailstock tap-holder, 3.91
Tank, heavy, repair of, 12.01
Tap drill chart, 6.12
Tape, double-coated, to hold workpieces, 3.28, 7.05
Tape-proofing, NC, styrene test for, 8.60
Tapered arbors, removing from drill chuck, 1.36
Tapered-box housing, 8.12
Taper pin, spring-loaded, 11.05
Tapers
  attachment for, 3.92
  checking on the level, 10.19
  grinding, 4.03
  setup for turning, 3.01
Tap-guide, spring-loaded, 6.21
Tap-holder, 3.91, 6.07
Tapped holes, blind, removal of chips from, 6.17
Tapping, of large drums, 1.37
Taps
  adapter sleeve to align, 6.01
  chucking, 6.03
  modified, for hand-tapping, 6.16
  notched, 6.13
  trepanning broken, 6.24

T-back square, 10.55
Teleprinters, paper tray for, 13.23
Tempering, of oil-hardening tool steel, 13.32
Templates, see-through, 10.42
Thin disks
  chucking, 3.21, 3.66
  Woodruff key, to clamp, 3.101
Thin stock
  punching with drillpress, 1.14
  welding to heavy sections, 11.29
Thread-cutting, reverse, 6.19
Threaded workpieces
  cross-drilling of, 1.26
  gripping, 3.42
  turning, 3.65
Threading
  boring bar, 6.02
  holder, 6.10, 6.11
  lathe attachment for, 3.47
Thread pitch, 6.04
Threads
  milled on lathe, 6.22
  split nut to gage, 10.51
Thread-tool, 6.08, 6.15, 6.18, 6.25
Three-bit flycutter, 2.61
Three-flute tools, measuring of, 10.20
Three-jaw chuck, eccentric turning in, 3.32, 3.61
T-nuts, 2.40
Toolbox, Kennedy, adding drawer to, 13.02
Toolholders
  floating, 3.58
  Morse-taper, 3.54
  support for, 3.88
  to control depth in centerdrilling, 3.95
Toolholding, collets for, 3.23
Toolroom lathe, convert to tracer lathe, 3.03
Tool-stand, for milling-machine shanks, 2.62
Tool-storage block, 13.04
Toothpaste, for lapping, 13.47
Torch jobs, modified table and fixtures for, 5.07
Tracer lathe, converted from toolroom lathe, 3.03
Transfer punch, 9.20, 10.53
Trigger stops, 8.64
T-slot clamps, 2.53
T-slot covers, 13.37
T-slot stops, 2.49, 2.64
Tubular workpieces

Index

supports for, 3.64
turning of, 3.45
Turning
eccentric, 3.17, 3.71, 7.08
multidiameter, stop extenders for, 3.87
off-center, 3.99
of small diameters, 3.39
of threaded workpieces, 3.65
of tubular workpieces, 3.45
plastic, 3.94
precision, 3.93
single setup for, 3.75
small stock, 3.97
split expansion arbor for, 3.82
tools, stacked, 3.84
tubular workpieces, 3.45
Turret-lathe
ball-turning attachment, 3.09
bar-puller attachment, 3.10, 3.11
Twist drills, polishing flutes of, 1.32
Two-legged spider, 3.98

Valve indicators, 12.07
V-blocks, 10.13, 10.25, 10.59
made from angle plates, 10.02
shimming to level stepped shafts, 10.44
V-bottom bushing, used to center rough castings, 1.41
Vertical mill
copy-milling cams on, 2.12
drillpress mounted on, 1.21
edge stop for, 3.33
indicator adapter for, 2.28
spacers for depth adjustment, 2.51
used as lathe, 3.79
used for drilling deep holes, 1.13

Vertical spindle, for surface grinders, 4.25
Vise jaws
extending, 7.24
for gang-milling, 2.07
holding parallels in, 13.15
universal, 2.65
Vise, machine, setup plate for, 9.17
Vise stop, double-jointed, 7.07
Vise, to hold odd-shaped workpieces, 9.14

Washer-like components, 8.11
Washer, with prongs, 11.28
Wedge, handle for, 1.20
Welding
bandsaw blades, 5.14
projection, of aluminum, 11.24
thin-walled workpieces, to heavy sections, 11.29
wire, handle for, 11.16
Weld, overhead, at floor level, 11.18
Welds, sealant for, 11.23
Winding, special springs, 3.100
Wire
cutting short lengths of, 5.12
straightener, 13.36
Wire hooks, 8.36
Woodruff cutter, for ID grooving, 3.44
Woodruff key, to clamp thin disks, 3.101
Woodruff-style keyways, flycut, 7.09
Worm-wheels, 13.45
Wrenches
Allen, 7.06
calibrating, 11.02
drawbar, as soft hammer, 2.18
strap, 11.25

1 Drilling

1.01 Adjustable power feed for a manual drill

Our shop does many production drilling jobs. To reduce operator fatigue and to boost output, we added a power drive to feed the drill down and retract it. The attachment is suitable for relatively shallow holes, either blind or through.

The device is mounted on a bracket secured to the column of the drill press. The sketch illustrates the mechanical linkage, but to avoid confusion it does not show the geared-head motor that is also mounted on the bracket. The driving disk is mounted on the shaft of this motor, which is a universal (ac/dc) type so that an inexpensive electronic speed control can be used to regulate its rpm.

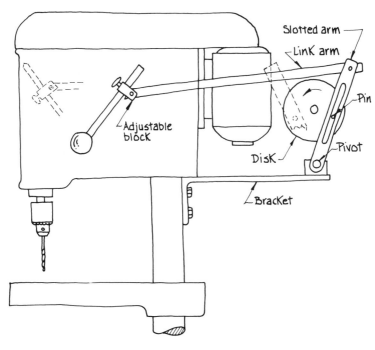

As the driving disk rotates from the position shown in solid lines to that shown in dotted lines, the drill is fed downward (note that this requires approximately 250° of rotation). As the disk continues to rotate (about 110°) the drill is retracted at a speed approximately double the feedrate.

Feed distance is adjusted by the position of the adjustable block on the drill's feed handle—the farther down toward the handle's center of rotation, the greater the stroke length. Feedrate is adjusted with the electronic speed control. Total cycle time, which is adjusted by the combination of feedrate and total stroke length, should be long enough to permit removal of a completed workpiece and replacement with the next one.

The device is especially useful for work that can be hand-held in a nest or against stops in a fixture clamped to the table.

CLINT MCLAUGHLIN, *Jamaica, NY*

1.02 Air-bearing base for heavy drilling fixtures

The sketch illustrates a setup we have used to drill a short production run of large castings in a heavy fixture. The operator actuates the air valve with his foot, which lifts the base to which the fixture is mounted and allows almost effortless repositioning of the heavy load from one spindle to the other. Stops affixed to the table provide for quick and accurate positioning.

A secondary benefit of the system is that the air pressure also clears chips from the table.

M K HAKIS, *Homer, Mich*

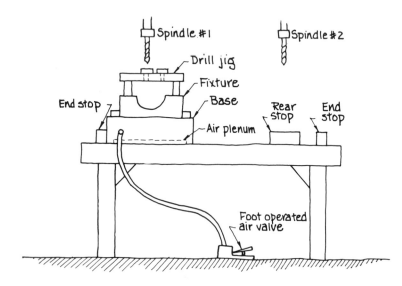

1.03 Attachment improves drill accuracy

In the machining of various parts it's often necessary to produce holes with tolerances on location, size, squareness with the surface, or other requirements that would typically necessitate the construction of a drill jig or fixture with guide bushings to ensure successful drilling or reaming on a standard drillpress. And if the part quantity is too small to justify the time and cost of making a drill jig, the holes may have to be produced on a more expensive machine by a highly skilled operator, or involve operations such as layout, stepping off, center drilling, redrilling, or boring.

Some of these problems can be overcome with two attachments we have added to a drillpress: one on the column, and one on the table.

On the column is a bushing-plate holder, fabricated as shown, and bored for a slip fit on the column, split in the rear, and screw-clamped securely in the desired position. Attached to this is an interchangeable plate (held by screws) to hold a drill-guide bushing precisely in line with the spindle. Each of four plates has a standard press-fit liner bushing with a different ID—5/16 in., 1/2 in., 3/4 in., and 1 in.—to permit the use of standard slip-renewable bushings for drills from No. 69 to 25/64 in.

On the table is a locating device, only one type of which is shown fastened to the table on the drawing. The one illustrated is a 90° "L" that is accurately positioned relative to the spindle by using a precision pin or rod inserted in the

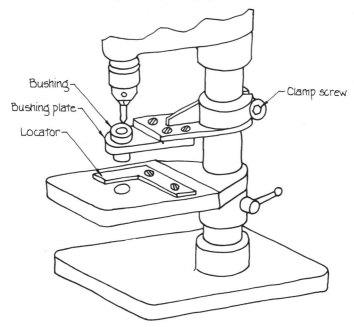

bushing. In our case, the two legs of the L are located exactly 4.000 in. from the bushing centerline. Various fairly obvious methods can be used to position a vise from this fixed locator. End, side, and solid jaw of any vise to be used should be square, of course, and we also measure the thickness of the solid jaw and stamp this offset dimension on it to simplify making new setups. We also use a workstop attached to the left end of the fixed jaw to ensure repeatable positioning of workpieces in the vise.

Using these attachments permits drilling to within 0.001 in. of desired location, and the holes are rounder, straighter, and of better finish than possible without a guide bushing. Also, if the bushing is positioned close to the work, holes can be drilled into angular or irregular surfaces without the drill walking.

I have also used this setup for punching holes in shim stock, paper, cork, rubber, and other thin materials used for gaskets, etc. A piece of steel plate is secured to the table and is drilled to the desired size. The twist drill is then replaced by a drill blank or pin of the same diameter that has been ground flat on the end. The "punch and die" are in alignment, and the drillpress is then used as an arbor press.

FRANCIS J GRADY, *Reading, Pa*

1.04 Band iron for hole layout

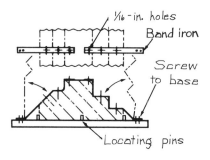

Short-run jobs with fractional-toleranced hole locations can be produced by the following method, which eliminates the need for both tedious layout of each workpiece and conventional high-cost drilling fixtures.

After a prototype part has been fabricated, mount, bend, and mark the required hole locations on strips of band (baling) iron, as shown in the drawing. Drill 1/16-in. holes at these marked locations on the band iron and mount it on the base. The workpiece should be nested between stops or over pins, and the run of parts is ready to be produced.

The flexible bands can be held against the part, and the hole locations are transferred by centerpunching through the 1/16-in. holes. For some jobs, several strips of band iron may be required, and they're simply pushed aside for unloading and reloading each new workpiece.

WILLIAM SLAMER, *Menomonee Falls, Wis*

1.05 Center-drill both ends

The drawing shows a setup for simultaneously center-drilling both ends of rods in a single operation. The setup is built around a heavy rod, threaded for its full length, which is secured to the drill-press column by plates at the top and bottom. Between these plates are a heavy-duty electric drill and a pair of V-notched guide plates for the workpiece. These guide plates should have oversize or slotted mounting holes for adjusting the alignment of workpieces, and should be case-hardened for longer production runs. Both center drills are mounted in commercial rotating drill holders that control drilling depth.

Use of the setup is simple and obvious. Just hold the work manually in the V-notches and feed the drill-press spindle to drill both ends. Because the two tools oppose each other, it's very easy to hold the shaft against rotation.

ERNEST JONES, *Yorktown Heights, NY*

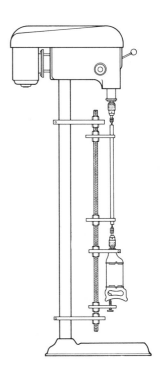

1.06 Centered depth-stop doesn't jolt sideways

Our shop uses a number of light drill presses for high-volume drilling. Typically, these machines all have depth stops alongside the quill housing—and when the stop-nuts hit the limit fork, the quill is pulled to the side.

To prevent any problems resulting from this sidewards pull, we made a special stop that mounts directly on the spindle. This consists of two simple parts: (1) A short length of bar stock bored to a slip fit on the spindle, taper-threaded with a pipe die on one end of the OD and knurled at the other, and slotted axially with a saw from the threaded end, and (2) a knurled pipe coupling cut in half to make a pair of locking nuts for two drill presses.

The sleeve is then slipped over the spindle

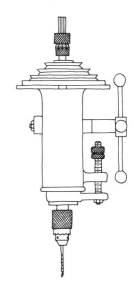

and is clamped by the nut wherever required for the job at hand.

The drill-press spindles are either splined or keyed and are driven by a cone pulley in which the spindle must be able to slide, which is why we decided not to use an even simpler collar with a set-screw that would probably raise a burr on the spindle. Spindle, stop, and pulley all rotate together, but because the pulley is a diecasting we also put a steel washer on top of the pulley.

CLINT McLAUGHLIN, *Jamaica, NY*

1.07 Chucking tiny drills

When the jaws of my drill chuck won't close enough to grip a very small drill, I wrap soft wire evenly around the tool's shank to increase its diameter for secure chucking. The trick prevents a lot of frustration.

R E JOHNSON, *Burnsville, Minn*

1.08 Cone simplifies shaft centering

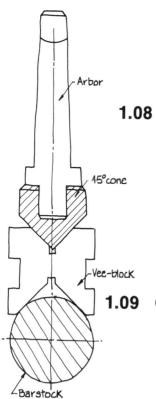

When it's necessary to drill a cross-hole through a shaft, a simple positioning method is to use a 45° cone-shaped plug in the drill chuck to center a V-block straddling the work. This effectively eliminates the disaster we all have sometimes of not drilling through center.

FRED STAUDENMAIER, *Bedford, Quebec, Canada*

1.09 Creating a slow-speed drillpress

Occasionally a shop has to fly-cut a few large-diameter holes in some sheet metal. That's one time when standard drillpresses typically don't have slow enough spindle speeds to do the job satisfactorily and safely.

Drilling

Not wanting to tie up a jig borer or a milling machine for such a simple operation, I recently came up with the following solution:

A 1/2-in.-capacity electric hand-drill is simply chucked on the unused bottom end of the drillpress motor shaft. (Most fractional-horsepower motors have double shafts projecting from both ends.) Handle of the electric hand-drill is placed against the drillpress column to take the torque and prevent the hand-drill from turning. The drillpress motor is left turned off, and the hand-drill is turned on. The combination of slow hand-drill spindle speed and the speed reduction of the drillpress's cone pulleys will result in a safe and efficient speed for the flycutter—generally about 100 rpm.

It's true that the rotation direction will be opposite of normal, but this is easily solved by reversing the flycutter bit so that it cuts in the opposite direction.

The whole procedure is as quick and simple as tightening the hand-drill chuck onto the drillpress motor shaft, and turning on the hand-drill instead. No modification is necessary to either the drillpress or the hand-drill, and the rig can be removed in seconds to return the drillpress to normal operation.

JAMES R RIDDLE, *Rochester, Pa*

1.10 Deburring thin work

A quick and efficient method for removing burrs around drilled holes in thin materials involves a backup block of polyurethane foam about 3/4 in. thick. The work is laid on this, and a standard single-flute countersink is chucked in the spindle of a manual drillpress. The operator controls the cutting force by letting up or depressing the foam

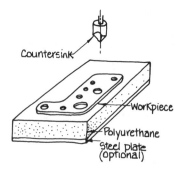

with more or less hand pressure on the workpiece.

A wide range of hole sizes can be accommodated in one setup, and either side of the work can be done without concern for any burrs still existing on the opposite side. The cutting tool does minimal damage to the foam surface, which allows many parts to be deburred on the same piece of urethane. The foam piece may be glued or pinned to a steel plate to facilitate positioning under the tool.

IVAN J GARSHELIS, *Pittsfield, Mass*

1.11 Drill pilots counterbore

Use of a short, standard drill inserted in a counterbore converts the tool into a step drill and combines two operations into one. Care should be taken and chips should be removed frequently when counterboring to depth.

ANTHONY J BABBARO, *Ithaca, NY*

1.12 Drill press table gets a lift

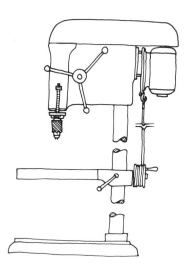

A variety of arrangements have been devised by machinists to help them in raising the tables of drill presses that were built without some kind of elevating feature. The drawing depicts a fairly simple device we built to accomplish this.

A hole was drilled and tapped at the rear of the table for a shoulder bolt, and on this was mounted a flanged aluminum pulley that we turned for the job. Added to the pulley was a crank handle, also attached with a shoulder bolt. A length of 1/8-in. flexible steel cable is secured through one flange of the pulley, and the upper end is attached by an S-hook through a hole drilled in the motor mount, as illustrated.

To raise or lower the table, you just tension the cable with the crank, loosen the table clamp, crank the table into position, and retighten the clamp.
   CLINT MCLAUGHLIN, *Jamaica, NY*

1.13 Drilling deep holes on a vertical mill

The project was to drill four 1/2-in. holes edgewise through a 14-in.-square block of aluminum 1 1/4 in. thick. Only six pieces were required. No gundrilling

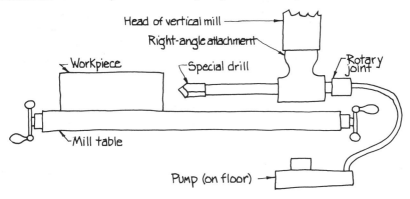

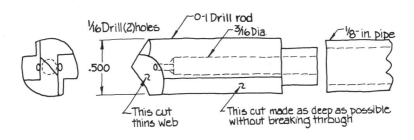

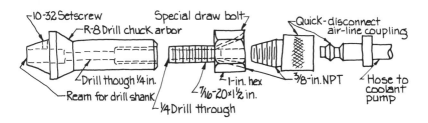

equipment was available. And special tooling or attachments were too costly and required too much lead-time. I decided to build my own setup, which took less than 4 hr from conception to completion.

The first decision was to do the job on a Bridgeport mill, using a right-angle attachment to obtain the necessary stroke of more than 14 in. The first sketch shows the basic setup.

The deep-hole drill shown in the second drawing was produced in about a half hour—the 0.500-in.-dia tip made from oil-hardening drill rod and attached to a shank made of 1/8-in. (nominal) pipe, which has an actual OD of 0.405 in.

To get coolant to the drill point, an R-8 drill-chuck arbor was modified (see third sketch) by adding a quick-disconnect airline coupler to act as a rotary coolant inductor. Rotary joints typically cost about $50; the coupler is readily available in hardware or auto-supply stores for about $4. The coolant pump was a Little Giant Model 1-Y submersible pump (intended for garden fountains and such uses). This cost $35.79, and operates at 2–3 psi. The coolant tank was a bucket.

Because overflow onto the floor could not be prevented, plain water was used as a coolant with only enough additive to prevent machine rusting. I've found that water works nearly as well as kerosene when machining aluminum, provided the supply is generous.

The spindle was carefully trammed with an indicator to ensure squareness with the table and parallelism of the right-angle adapter with the table.

Drill bushings were not used to start the holes. Instead, each hole was started with a center drill, drill, and reaming operation with an end-mill about 0.002 to 0.003 in. undersize. Reaming was done with an end-mill to ensure that the starting hole, which was about 1 in. deep, was perfectly true with the spindle. A standard reamer, of course, would have merely followed the drilled hole.

The deep-drilling operation was then done in a single pass at a spindle speed of about 600 rpm, with table-feed by hand. Time per hole was roughly 15–20 min. Runout in all cases was less than 0.010 in., and maximum oversize was 0.005 in. A bit more care would probably have produced more accurate results, but this was easily adequate for the intended purpose.

ROBERT B SEROZYNSKY, *Saugus, Mass*

1.14 Drillpress punches thin stock

When there are just a few, relatively small, round holes to be punched in thin sheet stock, it can be done without elaborate tooling. Here's a method I have used for punching round holes in stainless steel 0.020 in. thick or less.

Select a drill of the same diameter as the hole to be punched.

Use it to drill a hole at least 1/2 to 1 diameter deep in any kind of stock that presents a flat upper surface. (I use carbon-or stainless-steel.)

Remove the drill from the chuck and hollow-grind the butt end of the shank with a small grinding wheel.

Replace the drill upside down in the chuck of the drill press and you're ready to punch.

I've found that for thin stock it just takes a slight bump on the drill press feed lever to do the trick with no wear or damage to the machine. If the die fills up with punched out disks, simply drill another hole. It is a good idea, of course, to clamp the die to the machine table and periodically check its alignment.

BRYAN J SEEGERS, *Gilbert, Ariz*

1.15 Flip-top drill jig fits on machine vise

For close-tolerance drilling and reaming, you can add a simple drill jig to most standard machine vises, as shown in the illustration. To guarantee repeatability, this should be attached to the fixed jaw of the vise, and if a hold-down is necessary the front end can be slotted for a clamp screw in the movable jaw (which you'll have to drill and tap, of course). Hinging the jig on shoulder screws allows quick interchangeability. If the job requires through holes, the work can be blocked up with a soft filler piece to prevent damage to the vise when the drill breaks through.

JOHN R MAKI, *Danvers, Mass*

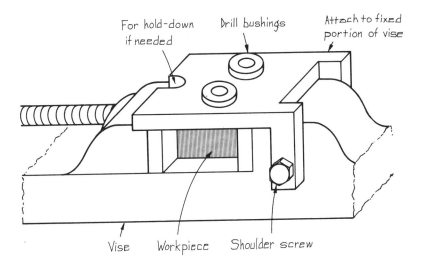

1.16 Flycutter adjusts from 3 1/4-in. to 8-in. dia

The photo and drawings illustrate a flycutter made for cutting circular holes in plastics, wood, and nonferrous metals, which has worked better than any commercial flycutter I have ever used.

The three basic parts shown on the drawing are reasonably easy to make. The given dimensions are not critical in most cases, though some care should be taken to ensure good fits. With the dimensions shown, the flycutter is adjustable to cover a cutting-diameter range of about 3 1/4 in. to 8 in.

I have also made a sturdier version for heavy-duty use in ferrous metals. It uses the same principles, but the sizes have been increased somewhat.

JOHN B HENDERSON, *Quakertown, Pa*

1.17 Geared gadget drills holes in tight corners

We had to drill two holes in a retrofit cover, but this seemingly simple task created a problem because the holes were at the ends of a long recess and located too close to the bottom and side of this recess to permit the use of either a conventional hand drill or ordinary machine tool (see smaller sketch).

The problem was solved with a gadget using a pair of mating spur gears. The smaller of the two was fixed to the drill bit with a setscrew, and the other to a pilot shaft. A guide (long enough to serve as a handle) and a drill jig were made, both with holes spaced to equal the spur-gear center-to-center distance. Two pilot holes were drilled in the jig so that it would serve for both the left-hand and right-hand operations.

The end of the drill shank was rounded and polished in a lathe, and a thick disk was fastened to the pilot in a position that would bear on the rounded end of the drill to serve as a thrust washer.

With the drill jig clamped in place, the drill and pilot were inserted and the holes were drilled.

If a reversing drill had been available, a standard right-hand twist drill could have been used. Lacking that, a left-hand drill bit would have served. Since we had neither, a right-hand twist drill was reground to cut with left-hand rotation. Because only a few covers had to be drilled, unhardened cold-rolled steel was used for the jig and guide.

ARTHUR DRUMMOND JR, *Walworth, NY*

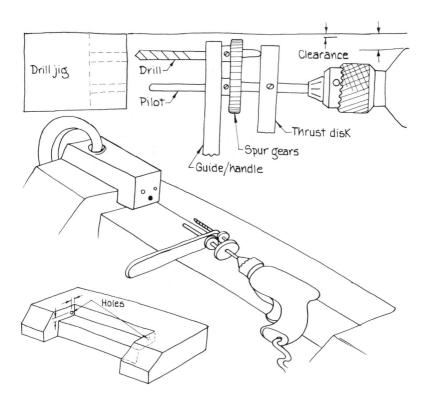

1.18 Gravity is helpful for angled machining

We fasten a standard protractor to the end of a long shaft as an aid in locating holes at specified angles as they are located axially along the length of the shaft for drilling.

The protractor is affixed to an end cap that facilitates attaching the gadget to the shaft, and a weight attached to the loose arm of the protractor makes it swing like a plumb bob to tell the operator its angular orientation.

Ernest J Goulet, Middletown, Conn

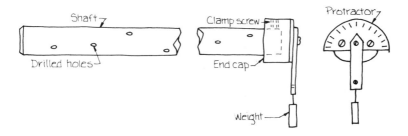

1.19 A handy drill rack

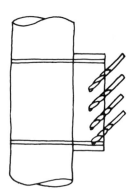

The best place to keep your most frequently used drills is right on your drillpress column. Just take a block of wood, drill it at 45° in a convenient pattern of holes for your most used drills, mark the sizes, and strap it to the drillpress column in a location that will not interfere with setup and operation of the machine. It will save a lot of time, money, frayed nerves, and spoiled work.

FEDERICO STRASSER, *Santiago, Chile*

1.20 A handy handle for a knockout wedge

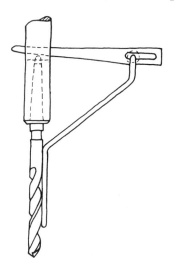

Removing a taper-shank tool from a machine spindle is really a three-handed job—and not many machinists are that well equipped. To prevent the tool from dropping onto the workpiece or the machine table or the operator's foot, we altered the wedge as shown in the drawing. A slot is milled in the wedge; the rod is bent as shown and is retained in the slot by a washer secured on the end.

To use the modified wedge, it's positioned as shown and one hand is used to grasp both the drill and the rod while the other is used to tap the wedge. The wedge will not bounce and fly, and the drill will not fall.

CLINT MCLAUGHLIN, *Jamaica, NY*

1.21 'Hang' a drillpress column on miller

We have a delicate, precision drill press that we often mount in a vertical mill (as shown) for precise location of small holes. And it gives an added bonus of increased throat depth for using the drill on larger workpieces.

All that's needed is a new stub-length column turned to the proper diameter and suspended "upside down" on the rear end of the mill's ram. Care should be taken that the post is square with the machine table, and the drill spindle should be checked by mounting an indicator in the chuck and sweeping it in a circle around the table.

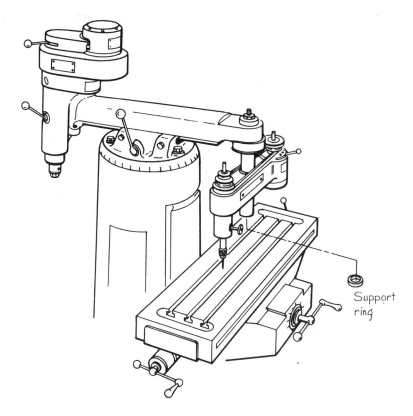

Support ring

For safety, a support ring should be affixed to the column below the drilling head so it can't slip off the bottom.

When the unit is needed, it's a simple matter to swing the ram around on its turret to position the drill spindle over the work area.

JOHN URBAS, *Cannon Tool Co, Canonsburg, Pa*

1.22 'Hoop skirt' eliminates coolant shower

We solved some NC drilling problems with indexable-insert drilling tools of the Kennametal/Metcut type. But that brought a new problem: a huge spray of coolant (which is fed through the shank) that not only showered the whole area but also quickly drained the machine's coolant tank and caused a lack of pressure for the drill.

To keep the cutting fluid inside the machine, we made a "hoop skirt" shroud of fairly heavy, transparent plastic sheeting. This is wrapped around the spindle housing and held there by a large hose clamp. The "hem" of the skirt is fanned out and attached with the duct tape to a steel-rod hoop welded at the ends.

It looks something like a lamp-shade, but it eliminates the shower and the loss of coolant from the tank. And it's flexible enough so that there are no interference problems as the spindle advances and retracts or as the table repositions; the skirt is simply pushed aside.

The idea may not seem very elegant, but it permits us to push 2-in. holes through solid steel on a 5-hp machine that's only rated for a 1 1/4-in. twist drill.

JOE RICKWALDER, *Kenilworth, NJ*

1.23 How to cross-drill screws

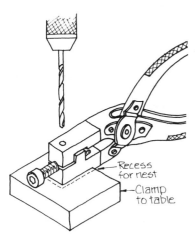

Cross-drilling of screws is a common shop operation that can be greatly facilitated with this simple tooling. The drill jig consists of a pair of parallel-jaw pliers with special tool-steel jaws welded in place. These jaws are drilled and tapped to accept the threaded workpiece, drilled to locate the cross-hole, and then hardened. The projecting lug on the bottom jaw accommodates a setscrew used as an adjustable depth stop.

To use the tool, the workpiece screw is placed in the open jaws and is turned in until it hits the stop screw as the jaws are closed. The bottom jaw is then placed in a nest clamped on the drillpress table to facilitate location under the spindle. And then the hole is drilled.

To remove any burrs that may have been generated, the screw head is then inserted into a rubber socket with a conically tapered hole, which is mounted on a small electric motor. This unscrews the part, which drops into a bin or box, and the edges of the hardened jaws deburr the hole.

CLINT MCLAUGHLIN, *Jamaica, NY*

1.24 It's a platform chuck

The platform chuck consists of an angle plate with an integral arbor for chucking purposes. Workpieces are positioned by a nest and secured with a strap clamp and a single capscrew. A single press-fit stop pin positions the strap for clamping and allows 90° rotation for loading and unloading.

The device is cheap, easy to make, and extremely useful for low-volume production or hard-to-clamp oddly shaped parts. It can be built as a dedicated tool with a specially shaped nest machined in it, or it can be made (as shown) to accept any number of removable adapters for different parts. The setup shown is for cross-drilling hubs.

Metal-filled epoxy can be used for making irregularly shaped nest.

DALE M GASH, Clayton, Ga

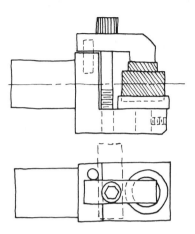

1.25 Jacking up a drillpress

Some drillpresses do not have a gear rack for raising and lowering the head. The way we do it in our shop is as follows: Put a suitable block of wood (or whatever else is handy) on the table under the spindle. Turn the travel-limiting stop nuts all the way up. Open the chuck all the way, so the force will be against the chuck body, not the jaws. Pull the handle down to put pressure against the block, and loosen the column clamp with your other hand. Now you can raise or lower the drill head with the handle. But don't forget to tighten the column clamp again before you let up the handle.

JAMES SERRATORE SR, Hatfield, Pa

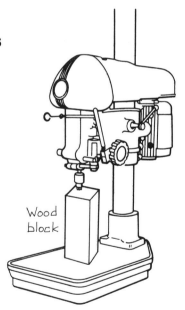

1.26 Jig for cross-drilling threaded parts

Cross-drilling screws is a common operation. The drill jigs we use for this incorporate a standard solid die to deburr the screw after cross-drilling.

The fixture is made of a steel block bored to accept the die, as shown in the sketch. A

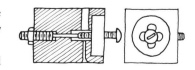

clearance hole is provided beyond this bore, followed by a smaller hole, the outer end of which is opened up and tapped for an adjustable positioning screw with a long pilot on it. An external locknut clamps this in position.

A hole in the top of the block guides the drill, although this can be modified for a drill bushing if many pieces are to be cross-drilled. A setscrew is added on the side of the block to secure the die, which we mount with the beveled side of the hole outward for easier entry of the screws to be drilled.

CLINT MCLAUGHLIN, *Jamaica, NY*

1.27 Locators for drill jigs

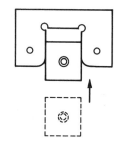

To speed the operation by facilitating manipulation, small drill jigs are often used without clamping them to the drill press table. Still more time can be saved by preparing a special locator plate of cold-rolled steel that is clamped to the table to provide a nest for the jig that instantly aligns it correctly with the drill spindle. The upper drawing shows such a locator for aligning a single hole; the two lower drawings show a locator used for aligning two holes.

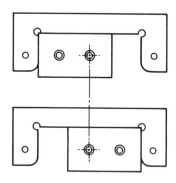

In making such locators, it's a good idea to radius or chamfer the entry side of the locator for easy entry of the jig. And if it's possible to make all of the drill jigs in a shop to a few standard dimensions, it's possible to gain these benefits with only a few locator plates.

FEDERICO STRASSER, *Santiago, Chile*

1.28 Mini drillpress mounts on larger machine

We have many heavy drillpresses in our shop, but we occasionally find it necessary to drill a few very small holes. For this work we made up a small

Drilling

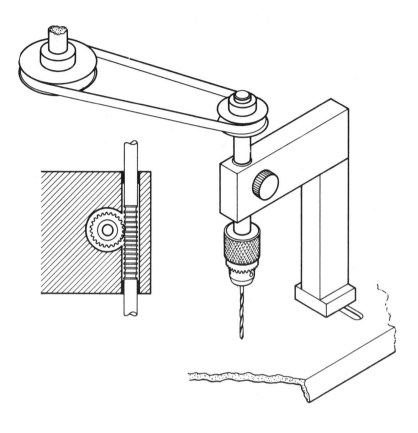

sensitive drillpress that mounts on the table of a large drillpress. The spindle is belt-driven from a pulley mounted on a short shaft held in the chuck of the large machine, with pulley diameters selected to give a speed increase. A round leather or plastic belt is used to avoid any problems of pulley misalignment due to feeding the small drill.

The "base" and "column" of the mini drill are welded up from rectangular barstock, and a bolt up through the bottom secures it to the table. The spindle has a number of grooves turned in it, as shown, to make a sort of round rack for feeding it by means of a gear in a counterbored hole in the head. We just use a knob for feeding, though a more-conventional feed lever could easily be added. And we just hold the spindle up with the knob, which could also be easily accomplished by a coil spring over the spindle below the pulley.

CLINT MCLAUGHLIN, *Jamaica, NY*

1.29 Modified counterbore as countersink

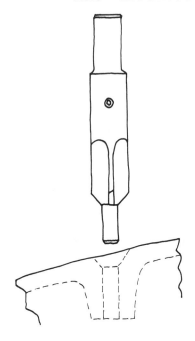

Machining some castings in our shop required drilling and countersinking a number of holes, one of which was not perpendicular to the surface (see sketch). Drilling was no problem, since this was done from the opposite side, where the hole was square with the surface. But countersinking an angled hole is difficult to do accurately because the lopsided initial contact forces the countersink "downhill."

We solved the problem simply by regrinding a counterbore to the angle required for the screw head. The c-bore pilot now keeps the tool perfectly in line with the drilled hole. The end of the tool was ground square across until the diameter at the end was equal to the pilot diameter. This eliminated the V-notch at the top of the pilot, which could have trapped chips.

CLINT McLAUGHLIN, *Jamaica, NY*

1.30 Part provides its own bolt-hole indexing

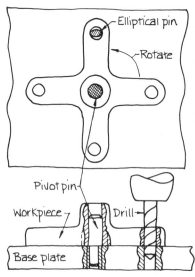

A simple method to position a workpiece for drilling and reaming a bolt-hole circle—or provide regular indexing for other machining operations—is shown in the accompanying sketch.

The job is done on a piece of tooling plate clamped in a vise or bolted to the table. This contains a shouldered pivot pin on which the workpiece can rotate and a shouldered elliptical pin the size of the bolt holes. The latter pin should be a slip-fit in the plate. The plate must be carefully located on the machine table so that the spindle center line is precisely the desired radial distance away from the pivot pin and exactly at the desired angular position from the elliptical pin (90° in the case illustrated).

After the first hole is drilled, the elliptical pin is inserted in the tooling plate and the workpiece is placed over both pins. This automatically positions the workpiece for drilling the second hole. And the sequence is repeated until all holes have been drilled.

Note that the technique is equally useful for producing bolt-hole patterns with any number of equally spaced holes. For a six-hole circle, the rotation angle would be 60°; for eight holes, 45°; and so forth.

JOHN R MAKI, *Beverly, Mass*

1.31 Piloted spade drill is chatter-free

Chatter problems with large spade drills can be eliminated by regrinding the blade as shown in the sketch. A pilot hole will not be needed to start this drill; a center hole is sufficient—and the spade drill will follow and not chatter while reaching the full diameter of the hole, even if the tool is extended at the end of a long boring bar.

MARTIN J MACKEY, *Parma, Ohio*

1.32 Polished flutes aid in chip elimination

If you polish the flutes of twist drills to about a 5- or 6-μin. finish with a Dumore or similar hand grinder, the chips will come out of the hole much more easily and faster. This reduces the required thrust on the drill, and it also prolongs the life of the tool.

MARTIN J MACKEY, *Parma, Ohio*

1.33 A quick cure for a sloppy drillpress spindle

A drillpress with excessive clearance between the quill and its housing can be a very frustrating tool to use. And some machines have no split-head adjustment to take out this slop. But they can be saved with this fix, which uses four small brass plugs snugged up with setscrews.

Remove the quill assembly and determine the upper and lower limits of its

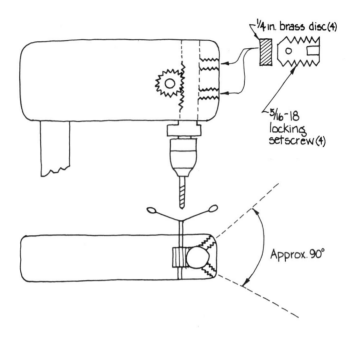

guiding bore. Drill and tap two 5/16-18 holes near the top and two more near the bottom of the bore (see sketch). Each pair of holes should have an included angle of about 90°. Clean up the debris, lubricate the bore, and reinstall the quill assembly.

Now slip a small brass disk (1/4 in. dia by about 3/16 in. thick) into each hole and follow it with a locking-type setscrew, such as the type with a nylon plug in the threads. The setscrews can now be used to restore minimum operating clearance and nominal drag.

CHAUNCEY VARNEY, *South Hero, Vt*

1.34 Rag ends drill chatter

To prevent a large drill from chattering as it enters a surface previously drilled with a small hole, just soak a shop rag in oil or coolant and put it on the point of the drill. The drill point will now cut to its full diameter without chatter.

MARTIN J MACKEY, *Parma, Ohio*

1.35 Recessing tool for drillpress

Here's a tool that will cut a recess or groove in the ID near the bottom of stepped or blind holes. It's made of tubular tool steel, tapered and split in three sections for part of its length, and with three cutting teeth of the desired form at the lower end. It should be hardened, of course.

An internal, loose-fitting mandrel is free to slide vertically, but is retained by being headed over at the top end or with a retaining ring or some other device. The bottom end of this should be turned to make a suitable pilot, above which should be a flat shoulder, and above that is a conical section as shown in the drawing.

In operation, the tool is lowered into the hole until it contacts the bottom or internal shoulder. Continued feeding pressure then forces the upper conical section into the bore of the recessing tool, expanding it like an ID-gripping collet so that it cuts the recess. Once the cut has been made and feed pressure is relaxed, the tool's spring action retracts the cutting teeth from the groove and the tool can be easily withdrawn clear of the hole ID.

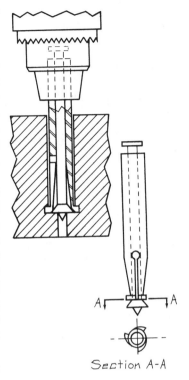

KEON GERROW, *Racine, Wis*

1.36 Removing tapered arbors

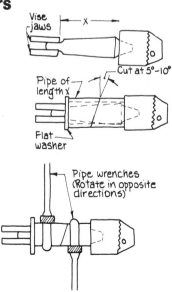

On occasion, a pressed-on tapered arbor must be removed from a drill chuck, and there just isn't enough shoulder on the arbor for chuck-removal wedges to work. The drawings show a quick and easy alternate method that uses a short length of pipe slightly larger in ID than the largest OD of the arbor.

Clamp the tanged end of the arbor in a vice, and cut the pipe slightly shorter than the distance between the vise jaws and the rear of the drill chuck. Cut the pipe in half at an angle of 5–10°. Now place the two pieces of pipe on the tapered arbor with the angled cuts mated, put a flat washer on the end as a bearing surface, and reclamp the tanged end in the vise.

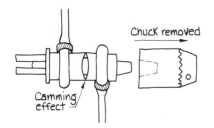

Finally, with a pair of pipe wrenches or Vise-Grip pliers, rotate the two sections of pipe in opposite directions. The camming effect of the angled surfaces will transmit a force between the vise jaws and the chuck to press the chuck off.

MIKE CATES, *Carlsbad, Calif*

1.37 Rocking fixture for work-positioning

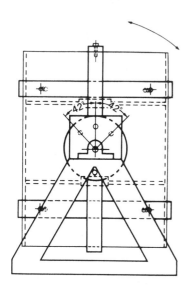

The fixture shown in the drawing greatly facilitates drilling and tapping of large drums with a radial drill. The drums must be drilled and tapped on both ends in an axis parallel to the centerline of the drum, and on one end in an axis 42° off the centerline.

We designed the fixture to pivot the drum and lock it in a vertical position with either end up, and also to lock one end at 42° off vertical in both directions. This allows us to drill and tap all holes simply by pivoting the drum to the various desired positions and locking it in place for machining.

To do the job without the fixture, we would have had to flip the work end-for-end with a hoist, and we would have had to use a portable drill in a holding fixture to drill and tap the angled holes. This would have added several minutes to the cycle time.

GARY R GREGG, *Shippensburg, Pa*

1.38 Sheetmetal drilling

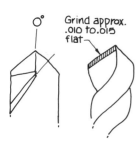

When drilling soft metals, such as aluminum, brass, and copper, or thin sheetmetal, a conventional twist drill has a tendency to "screw" itself into the job or lift the work. In jobs where this is a problem, a simple "fix" is to manually grind a slight flat, as shown, on the rake surface of the drill's cutting edge to reduce the positive rake from the helix angle to 0° or even slightly negative.

GARY KIPP, *Wickliffe, Ohio*

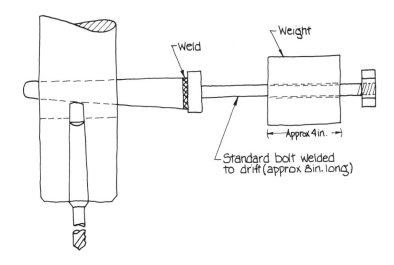

1.39 This impact drill drift saves a hand

Adding a slide-hammer to a drill drift permits taper-shank drills to be removed from a machine spindle with only two hands—never dropping either the drill or the drift. Just weld the head of a standard bolt (about 1/2-in. dia, 8 or 9 in. long) to a standard drill drift, slip on a 4-in. length of 1 1/2 or 2-in. barstock with a clearance hole through it, and add a nut at the end ot keep it all together. Use of the tool is pretty obvious.

WAYNE EVANS, *Poquoson, Va*

1.40 Use bushing bar to drill overlapping holes

Our tool design department prides itself in holding down the cost of slip spindle plates by trying to get as many hole circles or combinations as possible into each plate. Very often one hole will intersect another, causing problems.

I have overcome the problem by putting a drill bushing of the proper size in a separate bar that can be C-clamped or otherwise clamped to the spindle plate. I locate the bar on the work by positioning the table to true location and allowing the drill itself to position the bar (with bushing in place, of course) by hand-rotating the drill or using

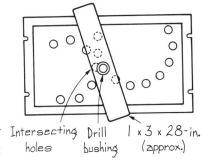

the jog cycle. When the location is established, the bushing bar is firmly clamped.

HERMAN KUBOWSKI, *Racine, Wis*

1.41 V-bottom bushing centers rough castings

When you get into a situation where you must maintain the spacing of two holes in a casting and it's also important that the second hole be central to the sides of the rough casting, this holding-plate and V-bottom bushing arrangement can speed and simplify the operation.

After the first hole has been produced, the part is placed over a locating pin on the fixture plate, which has been clamped to the machine table the precise distance (Y) away from the spindle centerline required. Next, the special bushing with its V-bottom is slid up the drill, the workpiece is swung under it, and the bushing is brought down onto the casting—contacting on both sides to center the work. With the bushing held firmly against the part, drilling is started, after which it's no longer necessary to maintain downward pressure on the bushing. The part can finally be removed by swinging it away from the drill.

JOHN R MAKI, *Beverly, Mass*

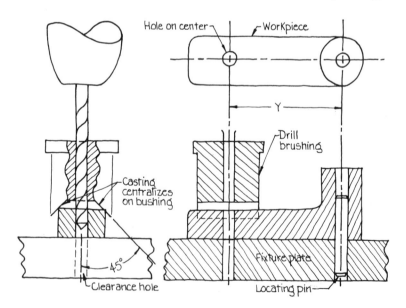

2 Milling and Boring

2.01 Adapter plate aligns opposing blind bores

This adapter plate makes a handy fixturing component for use on a small jig borer when you have a workpiece calling for bores on opposite sides that have to be in line within 0.001 in.

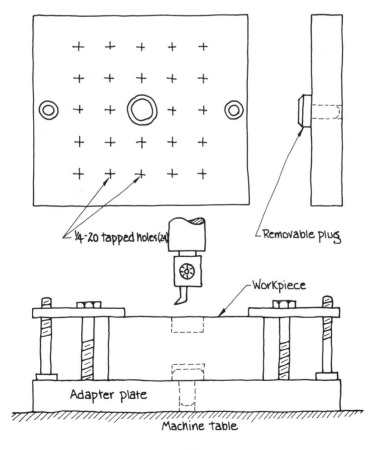

Although it could be made to any dimensions suited for your particular jobs, mine is made of a 6-in. x 7-in. piece of hot-rolled steel 5/8 in. thick, which was squared up and had both faces ground flat and parallel. The center hole is drilled and reamed to 3/8 in. diameter; a grid of 1/4-20 holes spaced 1 in. apart serves for part clamping; and two holes are drilled and counterbored on the centerline 1/2 in. in from the edges for 3/8-in. socket-head bolts to anchor the plate to the machine table.

For aligning opposed blind bores in a series of parts, one side of each is first bored conventionally. Then a plug is made up to fit the bores just produced on one end and the 3/8-in. reamed hole on the other. With a dial indicator in the spindle, the plug inserted in the adapter plate is zeroed under the spindle, and the table is locked in position. Now the workpieces are inverted over the plug and the opposite sides are bored—in line.

The plug can be reused for bores of the same size, or can be recycled for other jobs by carefully turning down to fit smaller bores. The plate, incidentally, can also be used in a lathe, using a four-jaw chuck for centering it.

JOHN URAM, *Cohoes, NY*

2.02 Adapter plate extends table's tooling area

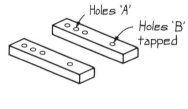

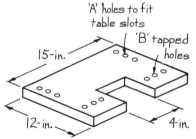

This adapter plate is for holding indexing heads, rotary tables, dividing heads, collet fixtures, and other such tooling attachments out over the edge of a milling-machine table. It's especially useful for holding long work vertically.

At first we just used a pair of rectangular bars drilled and tapped as shown in the upper drawing to fit the machine table and the attachment. From this evolved the more specialized plate—ours is 1 1/2 in. thick—shown in the lower drawing.

Holes marked 'A' are drilled to clear 1/2-13 socket-head capscrews and are counterbored for the heads to keep the working surface clear. Holes marked 'B' are drilled and tapped 1/2-13 for capscrews to secure the attachment desired. A number of different hole patterns can be drilled in a single adapter plate to suit a variety of attachments that may be used.

JOHN URBAS, *Cannon Tool Co, Canonsburg, Pa*

Milling and Boring

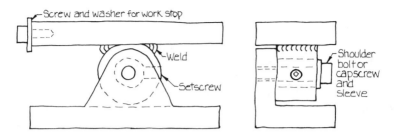

2.03 Adjustable angle-block fits in milling vise

Special angle blocks or wedges to tilt a workpiece in a milling machine vise at some particular angle have always been a bother to me, so in my spare time I picked up a few pieces of scrap and made an adjustable one. It works fine on a Bridgeport or similar mill.

Dimensions are certainly not critical, but the height should be no more than 2 in. because the depth of the standard Bridgeport vise is 2 1/8 in. Also, the narrower you make the assembly, the smaller will be the pieces you can clamp without shimming, although shimming is possible for very narrow work. The top piece should be about 3–4 in. long and about 1 in. or so in width.

For short production runs, the screw-and-washer work-stop at one end makes sure that all parts will be at the same height for milling operations.

Precise angles can be set with a protractor, and a variety of clamping schemes can be used in place of the simple setscrew shown on the drawing.

WAYNE GOODRICH, *Springvale, Me*

2.04 Adjustable pipe stops steady large parts

When machining fabrications on various machines in our shop, we would sometimes experience workpiece deflection or chatter. What was needed was

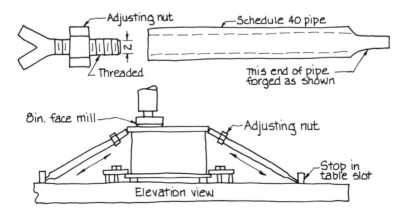

a fast and easy method of supporting these jobs for machining that would not add significantly to setup time.

The solution proved to be the use of adjustable pipe stops (see sketch), which were both easy to install and effective in supporting the work. At the bottom of the sketch is a typical example of a setup on a planer-mill table. Tightening the adjusting nut provides support to resist the cutting forces from an 8-in. face mill. And the pipe stops can be used on all types of machines.

ROBERT SZEMANSKI, *Pittsburgh*

2.05　Adjustable spacer for straddle milling

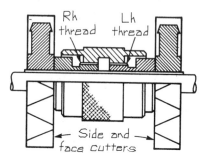

In straddle milling, we often find that the distance between the two cutters must be altered slightly—a few thousandths of an inch. On one such occasion, we decided to make an adjustable spacer, as illustrated, instead of using a conventional solid spacer or shim. Considerable operator time and effort are saved.

As the outer knurled sleeve is rotated, the inner threaded bushings move either closer together or farther apart—guided by the keyway in the arbor—varying the spacing of the cutters over an adjustment range of about 1/4 in. One cautionary note is that the spacer unit should be mounted in such a way that any loosening tendency of the outer sleeve would tend to cause the bushings to move outward.

W T CHAKRAVARTHY, *Madras, India*

2.06　Attaching attachments

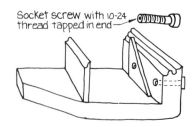

Needing a way to secure different stops and angles in a milling vise, I removed the socket-head screws holding the hardened jaw and drilled and tapped the ends of these with 10-24 threads. Then I replaced them in the vise and had ready tapped holes for any possible vise adaptation need.

The sketch shows how a 45° angle is held by this method.

The idea quickly caught on throughout our department, and now all of the vises are so modified.

JIM THORNTON, *Toledo, Ohio*

2.07 Backup vise jaw helps gang-milling

Gang-milling a stack of sheets clamped in a vise is a convenient way to produce a short run of identical "stampings," but the last couple of sheets in the stack almost always are distorted when the milling cutter comes through. To eliminate this distortion and the scrappage it causes, a special vise jaw can be made to back up the stack of sheets.

This jaw, which should be attached to the vise and which can also include an end stop for convenient positioning of the stack, can be made of a rectangular piece of mild steel of 1/4-in. or greater thickness. When milling the first stack, just continue the cut right on through the unhardened backup jaw. It will then serve for the rest of the lot you're producing, and it can be removed from the vise and stored for future runs of the same job.

JOHN R MAKI, *Danvers, Mass*

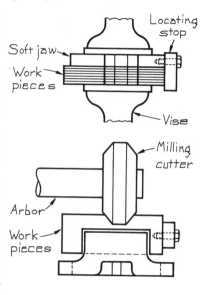

2.08 Ball clamp for odd shapes

Odd workpiece shapes or cast surfaces sometimes make it difficult to clamp a part on a milling-machine table with standard strap clamps. Special "ballfoot" clamps make the job easier.

Just center-drill both ends of the strap and then either cement or braze bearing balls into the recesses. A similar depression or hole in the support block helps to maintain a set position, and a set of spherical washers under the hold-down bolt will also help accommodate misalignments.

JOHN R MAKI, *Beverly, Mass*

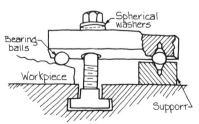

2.09 Ball detent holds loose quill clamp

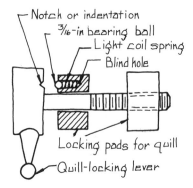

The quill-locking lever of a milling machine that's in constant use soon gets rather loose in its unlocked position. This allows the lever to drop by its own weight and put a drag on the quill, and this, at best, is an annoyance when using the machine for light drilling, or when using an indicator.

Some machinists solve the problem with a rubber band hooked over a nut, but the solution shown in the sketch is more workmanlike and is unlikely to develop any bugs.

Just drill a blind hole in the outer clamping pad that's a few thousandths larger than a 3/16-in. bearing ball. Insert a light spring and then the bearing ball. A mating notch or very shallow hole is then made in the inside surface of the lockscrew head. Position of this "dimple" has to be determined by preliminary assembly; it should be positioned so the ball will enter with the screw in its unlocked position.

DANIEL FORWARD, *Mountain View, Calif*

2.10 Center plug facilitates setup of rotary table

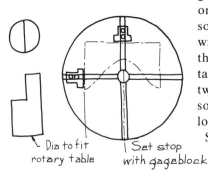

We've found that a simple—but carefully made—center plug with the upper end ground to center greatly facilitates setups on our rotary table, and enables us to do some of the same kinds of jobs we'd otherwise need a rotary cross-slide for. Fitting this plug into the center hole of the rotary table permits accurate setting of stops in two dimensions with precision gage blocks so that we can position work to a given location and radius.

STEPHEN G ROBY, *Flint Tool & Machine Co, Alexandria, Ind*

2.11 Centering a punch mark under the spindle

Here's a simple and accurate method for centering a transfer-punch mark under the spindle of a milling machine using a wiggler, a Last Word indicator, and a simply made 1/2-in.-square holder that links the two together as shown in the illustrations. The wiggler should be the type that incorporates a spring behind the ball end of the point.

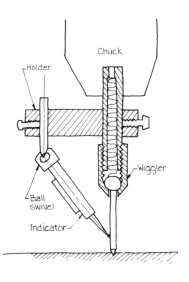

Insert the assembly in the spindle, and, with the punch mark fairly close to center, put the wiggler point into the punch mark. Make sure there's some spring pressure on the point. Then swivel the indicator over so its ball end is against the wiggler stem. Now, rotating the drill chuck manually, true up the dial readings by turning the table cranks. When the indicator shows a constant reading all the way around, the spot will be dead true with the spindle.

KENNETH JUDSON, *Waterbury, Conn*

2.12 Copy-milling cams on a vertical mill

Several times a year we have machined thousands of plastic cams for a certain customer on our NC machining centers. Machining takes only a few seconds each, but setup and inspection time is considerable. Normally, this is no problem because of the large lot size.

But then this good customer asks for a short run—200 pieces—and expects to pay the same price per cam as before. The NC equipment, of course, is all tied up on other work, and there's a natural reluctance to break down that job for 6–8 hr of setup and 2–3 hr of machining on the cams. And the shop has no tracer mills.

Here's one solution on a standard milling machine: Take a two- or four-flute end-mill (standard length or extra long) and make sure of its diameter. Then get a standard roller or ball bearing of the same diameter and grind the end of the end-mill down concentrically until it's a light press-fit into the bearing bore. Leave a little shoulder to make sure the outer race of the bearing clears the milling cutter body. This is then chucked in a horizontal attachment carefully mounted and indicated parallel to the machine table.

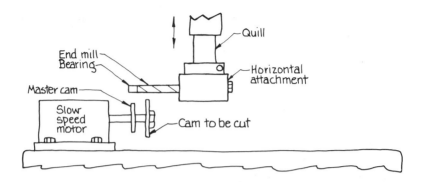

For holding and feeding the work, a low-speed motor, or gearmotor, or high-speed motor with speed control is used. This is fitted with a special shaft for holding both a master cam to be duplicated and the work blanks. Depending on the thickness of the cams and the length of the end mill, this shaft can be made to hold several workpieces for simultaneous milling. End of the shaft is tapped or threaded so that blanks can be clamped securely. The master cam should be permanently affixed, and the motor should be mounted on the mill table making certain that the shaft is parallel to the table surface. Shaft rotation should be opposite that of the cutting tool.

To mill cams, load the shaft with blanks, turn on the motor and the machine spindle, and (with the table clamped in position so that the master cam is directly under the bearing) feed the quill down slowly until the bearing is running on the master cam. Maintain light pressure on the quill, so the tool will follow the master cam's outline, and let the workpiece make at least two full revolutions before raising the quill.

If only a very few cams must be made, the expense of the slow-speed motor can be eliminated by substituting a manual cylindrical grinding fixture.

JOSEPH D JUHASZ, *Michigan City, Ind*

2.13 Cut keyways on a lathe

You can use a lathe to cut keyways in shafts simply by putting an end-mill in the spindle chuck or collet and mounting the shaft on the machine's cross-slide or compound perpendicular to the turning axis. Obviously the shaft must be carefully centered. Depth of cut can be set either with the compound (set up to feed toward the headstock) or the saddle. Then the cut is made by advancing the cross-feed. Using this setup, keyways can be cut either in the center of the shaft or at the end.

DAVID R LANDIS, *Chillicothe, Ohio*

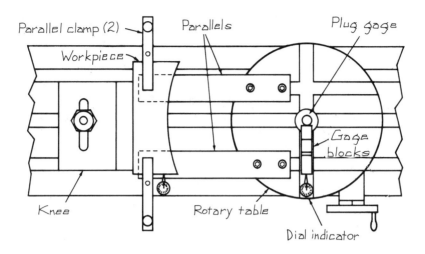

2.14 Cutting an arc when the radius is too big

Here's a setup problem for you. You've got to mill an inside arc with a radius larger than that of the rotary table on your machine. In fact, no part of the workpiece comes even near the rotary table. There are parts like that, such as a tracer-lathe templet recently needed in our shop.

The setup is shown on the drawing. First, a knee is trammed parallel with the in-out travel of the milling-machine table. Then the rotary table is positioned the correct distance away, using an inside micrometer and measuring from the knee to a plug gage in the center of the rotary table. (There's usually enough information to calculate this dimension.) And the rotary table is clamped in place.

If the center of the arc must be aligned with or offset by some specific distance, a dial indicator in the spindle (spindle not shown) is offset the proper amount from the plug gage with gage blocks and the dial is zeroed. Then the table is traversed to the workpiece, which is held against the knee, and the work is pushed against the indicator until it reads zero. At this point the work is clamped to the parallels mounted on the rotary table's T-slots. It's a good idea to traverse the table back to the gage blocks to make sure the indicator still reads zero. Finally the knee is removed and the radius, which was previously roughed out on a bandsaw leaving about 1/16 in. of machining stock, can be finish milled.

NILS G BRADLEY, *Westerly, RI*

2.15 Cutting keyways oversize

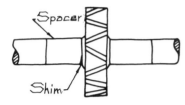

It is sometimes necessary to mill keyways a bit over the nominal dimension—for example, on parts to be heat-treated. A quick and easy way to do this with a nominal-size cutter is to place a shim between the cutter and the adjacent spacer, as illustrated. This makes the cutter wobble in the manner of a portable circular saw with dado washers installed. Width can be controlled quite accurately by using various thicknesses of shim material.

The method obviously works better with smaller cutters, and it can be a lifesaver when it's necessary to use a worn cutter.

TOM STAPLETON, *Plainview, Tex*

2.16 Deburr while machining

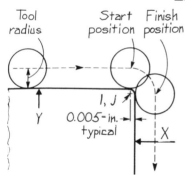

"Smooth all sharp edges, break all sharp corners, remove all burrs," is a common notation on many a blueprint. And many a 45° chamfer is specified for the same reason and no other. These are virtually machine shop traditions.

But when you're using a nontraditional numerically controlled machine tool, there's a fairly simple programming step you can take to achieve the same purpose in a somewhat unconventional way. Most modern controls provide a circular-interpolation capability, and it's quite easy to program the tool path for a small radius round a sharp corner, as shown in the drawing, setting I and J values at 0.005 in. inside the part outline. The technique can be used in milling or in turning, using K instead of J in the latter case, of course.

JOSEPH D JUHASZ, *Horsham, Pa*

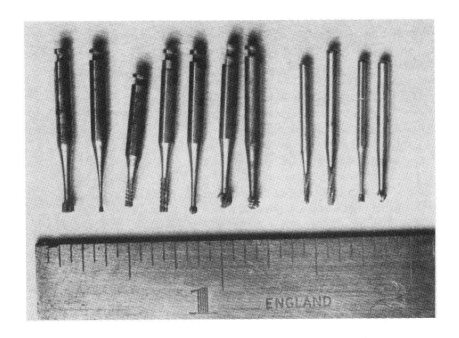

2.17 Dental burrs are miniature milling cutters

Keep your eyes open the next time you're sitting in the dentist's chair and you'll see some of the handiest little milling tools you could ever find for doing miniature tool and die work, modeling, prototype jobs, samples, and any number of other applications. These tools are dental burrs, sometimes called "drills," although they certainly aren't drills in the metalworking sense of the term.

The photo, which is reproduced exactly double actual size, shows a few of the many shapes and sizes available. They come in sizes from 1/64 in. to 1/8 in. The ones at left in the photo are high-speed steel, which comes with 3/32-in. shanks, and the ones at the right are carbide, which have 1/16-in. shanks. Split-sleeve adapters are easily made, if necessary, and the burrs can be used in drillpresses, hand grinders, lathes, milling machines, and other machine tools as well. They'll cut virtually any metal, wood, or rigid plastic material—even including modeling wax.

RUDOLPH K SCHMITT, *Whitestone, NY*

2.18 Drawbar wrench doubles as soft hammer

Anybody who has ever run a Bridgeport or similar milling machine knows the temptation, after loosening the drawbar with a 3/4-in. wrench to change collets, to hit the drawbar with the side of the wrench to break the collet loose. This isn't correct practice, of course; you're supposed to put down the wrench and use a small, soft-faced hammer. But the temptation is strong.

I solved this problem with a 1 1/4-in. length of 1-in.-dia brass barstock. I milled a 1/4-in. slot 5/8 in. deep in one end of this, inserted the wrench handle, drilled a 1/8-in. hole through both, and assembled it with a 1/8-in. Rollpin.

This makes a handy combination wrench and hammer—and it adds a good bit of life to the drawbar.

GARY KIPP, *Wickliffe, Ohio*

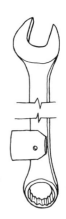

2.19 Enclosed tool tray keeps shanks clean

We use quick-change tooling with tapered shanks on our Bridgeports and

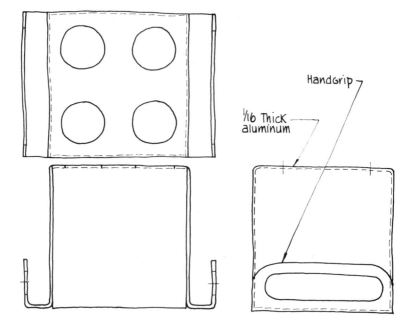

other milling machines. When the work requires three or four operations and several pieces have to be made, we bring all the toolholders with tools inserted to the machine table to save time. The problem was that the shanks couldn't be kept free of dirt and chips. A tiny speck on a tapered shank is enough to cause trouble, and the most irritating difficulty was that when we seated a shank in the spindle, we couldn't remove it later by hand. The quick-change tooling was no longer quick. And we didn't want to spend time wiping down each shank before it was used.

To keep the shanks clean and chip-free, we made some box-shaped tool trays. Made of 1/16-in. sheet aluminum (to prevent scratching), these are made of one-piece blanks bent into the configuration illustrated. With the open side of the box facing down, the top has four holes to suit the particular tapers we use, and the shanks are entirely enclosed. For easy handling, the tray even has grips.

ERNEST J GOULET, *Middletown, Conn*

2.20 Expand your mill vise capacity for flat work

Perhaps it's a corollary of Murphy's Law, but an awful lot of workpieces seem to be just over the capacity of a standard milling-machine vise. If the workpieces are basically flat, however, such as drill-jig plates, the simple vise modification shown in the sketch will just about double vise capacity.

A secondary vise jaw of 1/2- x 1/2-in. stock 6 in. long with two holes drilled and counterbored for Allen-head capscrews, then hardened and ground, is fitted into a snug slot milled across the vise as shown parallel to the fixed jaw. Matching holes are drilled and tapped at the bottom of this slot for the capscrews.

To use the new auxiliary jaw, the regular moving jaw will have to be

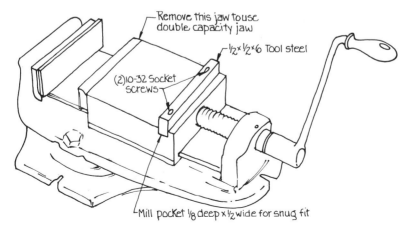

removed unless you also make a new fixed jaw that matches the total height of the new rear jaw.

WILLIAM HITCHEN, *Chicago, Ill*

2.21 Fixture mills radii without rotary table

Improvisation is the answer to milling radii when a rotary table is not available. A fixture can easily be made from a steel baseplate—its size governed by that of the workpiece—with a dowel pin pressed into it to serve as the pivot for a top plate arranged to clamp the work (see sketch).

With the baseplate secured in a vise or directly on the table of a vertical mill, the top plate and the workpiece can be fed past an end mill in the spindle by means of a sturdy rod screwed into a tapped hole in the edge of the top plate. Light cuts should be taken into the rotation of the cutter—up milling, in other words.

Many modifications are possible, depending on the specific nature of the job at hand. For example, stop pins can be pressed into the baseplate to limit the arc swung by the top plate.

The radius to be milled, of course, is measured from the center of the pivot pin in the setup sketched. In many cases, however, the workpiece will have a hole at the center of the radius, which will allow the work to be pivoted directly on the baseplate.

WILLIAM HITCHEN, *Chicago, Ill*

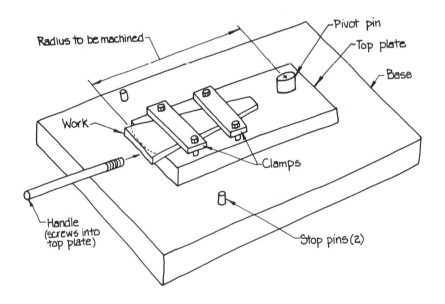

2.22 Flycut odd slots

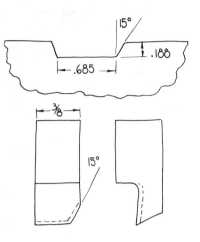

To machine odd-shaped slots of various widths, typified by the shape in the drawing, and without complex special cutters, I simply use an adjustable boring head mounting a flycutter made from the shank of a discarded end mill. The cutter shown was made on a surface grinder. Width of the slot is controlled simply by adjusting the slide on the boring head.

The system works quite well, and can also be used for straight-sided slots of nonstandard dimensions, or even dovetail slots.

ROBERT MESSINA, *Blairsville, Pa*

2.23 Hack off corners to save NC cutters

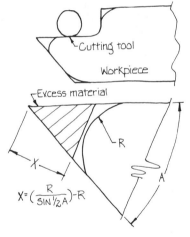

When corner-rounding in a numerically controlled machine, the cutter should not be overworked by removing excess material that could be removed in a less-costly way. As shown in the drawing, the "X" amount, or slightly less, can be economically hacked off to leave a more-uniform cut along the contour of the workpiece for the sharp, on-size, and probably expensive cutter in the NC machine's spindle.

The formula, incidentally, can be used for both acute and obtuse angles.

ERNEST J GOULET, *Middletown, Conn*

2.24 Horizontal mill turns spherical parts

Spherical parts ranging from ball-end tracer stylii to an 8-in.-dia hemispherical inspection master have been turned to demanding accuracy on a horizontal milling machine with the setup shown in the photo, which requires only simple tooling rather than special form tools or numerically controlled machines. The setup involves a simple, albeit rather tall, toolholding arrangement on top of a compound slide mounted on a rotary table.

Workholding is, as always, dependent on the work itself. For the large

hemisphere, the workpiece was fixtured to a faceplate mounted on the mill's horizontal spindle; for the smaller parts, a four-jaw lathe chuck was mounted.

Tool-height adjustments are made by moving the mill's knee up or down, while moving the cross-slides centers the tool point and determines the radius being turned. Moving the machine's saddle in and out positions the center of the radius in relation to the faceplate or chuck. And motion of the rotary table, of course, swings the tool through the radius.

KEVIN S BARNEY, *Monroe, Conn*

2.25 Horseshoe milling spacers

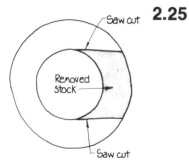

Take large washers—or make them. Saw a slot to the center hole wide enough to fit over the arbor of a milling machine. And grind both sides flat on a surface grinder to the required thickness. You now have milling spacers that can be changed quickly—without taking the arbor out of the mill.

MARTIN J MACKEY, *Parma, Ohio*

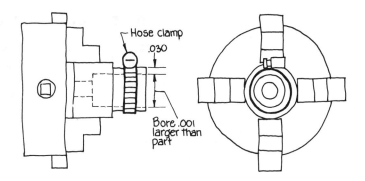

2.26 Hose clamp gently holds delicate work

Boring and reaming the ID of a thin-walled ceramic-coated workpiece was a tricky operation because it was so easy to chip or crack the brittle ceramic coat on the OD. This pot chuck did the difficult gripping job without any surface damage, however, and also provided an internal step for controlling the axial clamping position of the work. Because the chuck ID is only 0.001 in. larger than the workpiece OD, it also ensured close tolerances on concentricity.

Although this type of fixture is often split lengthwise like a collet, this was not done in this case. The combination of only 0.001-in. clearance and the 0.030-in. wall thickness allows the hose clamp to compress the pot chuck sufficiently to hold the work involved.

GARY A STEPHENS, *Taylorsville, Ky*

2.27 How to handle a heavy vise

Milling-machine vises are designed for clamping onto a T-slotted table and securing work under a spindle. They're not designed for handling—they're heavy and hard to grip—yet they frequently have to be moved around the shop.

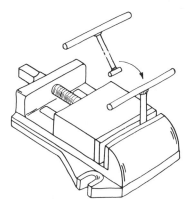

A simple handle welded up from three pieces of 1/2-in. barstock, as sketched, really helps. Just clamp the handle in the vise itself, and if the vise jaws don't close all the way a piece of scrap—even a wooden block—can be used as a filler to clamp the handle.

CLINT MCLAUGHLIN, *Jamaica, NY*

2.28 Indicator adapter for mills

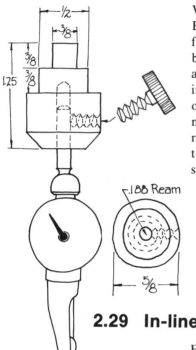

When working with a vertical mill such as a Bridgeport, interchanging collets back and forth to accommodate a dial indicator becomes a time-consuming nuisance. To avoid this, I made the simple adapter shown in the drawing so that it fits the indicator on one end and matches the shank diameter of my end-mills on the other. Now if I have to remove the end-mill so I can use the indicator in the spindle it's no longer necessary to swap collets, too.

RON STANWICK, *Englishtown, NJ*

2.29 In-line gearbox updates old VBM

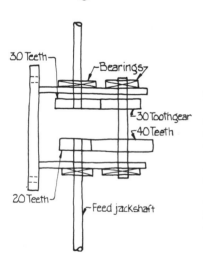

Feedrate selections are typically limited on most older vertical boring mills. You can often rev up the table speed sufficiently to make better use of modern cutting tools, but for most finishing applications the available feeds are generally too coarse.

But it really isn't too difficult to cut an in-line gearbox into the machine's feed jackshaft. The schematic drawing shows the principle of a 1:2 speed-reducer we have added to an older VBM, which reduces an 0.032-ipr feed to 0.016 ipr, for example. The basic idea is that each gearing stage must have a pair of gears with the same total number of teeth (30 + 30 = 60, and 40 + 20 = 60 in the example), and the overall ratio is the ratio of one stage multiplied by the ratio of the other (30/30 x 20/40 = 1/2).

WILLIAM H PATE JR, *Birmingham, Ala*

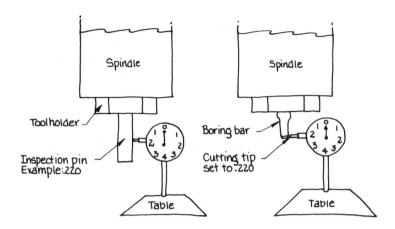

2.30 Inspection pins set boring tools

I have found a time-saving system for setting the diameter of adjustable boring tools in the spindle of a vertical CNC machining center.

Use an inspection pin that is 0.001 or 0.002 in. under the finish bore diameter. Put this pin in a toolholder in the spindle, and tram it to zero with a dial indicator set up on the machine's table. Now, leaving the dial indicator in place and being careful not to bump it, replace the inspection pin with the adjustable boring tool. Tram the point of the boring tool to zero, and it's now set for the 0.001 or 0.002 in. that the pin was undersize. The final setting is then routine and easily obtained.

KEVIN L MOORE, *Maple Heights, Ohio*

2.31 Machining square holes with round cutters

The sketches show two simple ways to machine a square hole with limited machines and standard tools. Sketch 1 illustrates the technique with an end mill, sketch two shows the method with a peripheral mill or Woodruff cutter. First drill a small hole on the center of the desired square hole. Next mill a slot

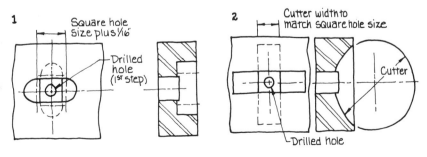

slightly more than half way through the work. Then turn the part over and mill another slot at right angles to the first.

The system doesn't apply in every case, but it's very useful for one-of-a-kind jobs that cannot justify purchase of a broach or other special tooling. We used the method to produce an anti-rotation hole for a sliding square bar.

KEN GENTRY, Houston, Tex

2.32 Mill table extended for splining long shaft

In times like these you just can't afford to turn down jobs because they're not ideally suited for setup on the particular machines in your shop. So you invent ways to handle them. An example was the 4-ft-long shaft that came to our shop requiring that a 6-flute spline be machined on it. It was just too long to be set up on the table of our vertical mill with the necessary index head and tailstock support.

So we extended the table of the milling machine, as shown in the sketch, by adding a 12-in. length of 1-in. x 4-in. steel to mount the index head at one end and a 24-in. length of the same material to support the tailstock at the other end of the table. All drilling and milling operations (for the key slots) were done on the extended machine, of course.

EDWARD LINDERMAN, Irons, Mich

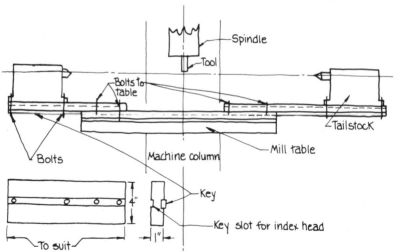

2.33 Milling a hex from a lathe turret

Here's a setup we use to cut a hexagon on the end of a brass rod. We do it in a small, hand screw machine because the workpiece is primarily a turning job.

Milling and Boring

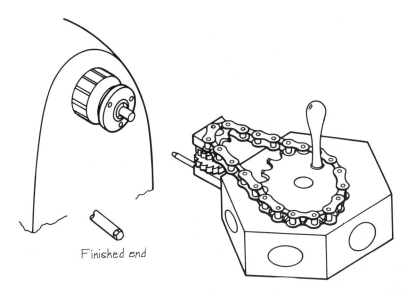

Finished end

We mounted a short shaft on top of the turret for a sprocket, to which a handle is fixed so it can be rotated by hand. A U-shaped block with a shank is mounted on one turret face. A short, vertical shaft runs through bushings in this block and carries two milling cutters and a spacer. A small sprocket on top of the shaft provides for driving it.

Indexing the work is taken care of by a steel pin projecting from the U-shaped block and a ring with three holes fixed to the spindle nose.

To mill the hex, the turret is advanced (with the machine's spindle stopped, of course) until the indexing pin enters the first hole in the index ring. Now the cutters are rotated while the turret is fed up to the stop. This is repeated twice more to complete the hex.

A slight curvature—equal to the radius of the milling cutters—is produced in the shoulder at the base of the hex, but this is not objectionable in our workpieces.

For larger work, it would also be possible to power the cutters with an electric drill or a small geared-head motor instead of using manual power.

ERNEST JONES, *Yorktown Heights, NY*

2.34 Milling crowned work with a flycutter

In the process of machining extrusion dies, I found that they worked better if the faces of the die block were crowned slightly. The face on one type of block is approximately 2 3/4 in. wide and 4 1/4 in. long, and the crown at the center is 0.010 in. higher than the edges. To machine this crown required that we

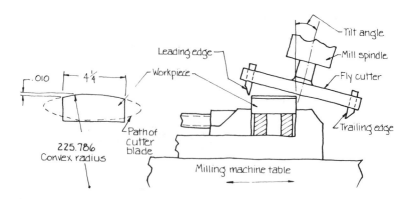

generate a radius of 225 3/4 in. on the faces of the die blocks, which we accomplished on a vertical milling machine set up as shown.

The block is held in a machine vise, and a flycutting tool is mounted in the spindle. Tilting the spindle as indicated causes the flycutter to describe an elliptical path relative to the machine's table-feed direction, which closely approximates a circular arc of the desired radius. A convex radius is produced on the surface of the block if it is fed under the leading edge of the rotating cutter; a concave surface would result if the block were fed under the trailing edge.

Since the method generates an elliptical surface, the nearest approach to a true circular radius results by using the largest possible cutter diameter and the smallest tilt angle. Large flycutters can be made economically by installing square-hole sleeves for lathe-toolbits in the desired location in a steel plate and attaching a shank to fit the milling machine spindle.

Relationship between the tilt angle, cutter radius, and workpiece radius is: cutter radius divided by work radius equals sine of the tilt angle.

FRANCIS J GRADY, *Reading, Pa*

2.35 Milling plates square

A frequent job in our shop is machining rather large plates square, and we've developed a simple and quick way to do it.

We make an aluminum stop block about as long as the milling machine table is wide. A step is milled along one bottom edge to provide cutter clearance over the table in a later operation, and holes are drilled for hold-down bolts. This is clamped to the table approximately square, and then a tru-

Milling and Boring

ing cut is taken with an end mill to get it perfectly square.

The workpiece is now butted against this and clamped with one edge slightly beyond the rear edge of the table, so the end mill won't cut the table, and that edge is milled. This newly milled edge is then butted against the stop, essentially as before, and the second edge is milled. The third and fourth sides are then done in the same way—producing a perfectly square workpiece with a minimum of setup. And the aluminum stop block can be used over and over.

WILLIAM RAHN, *Mahopac, NY*

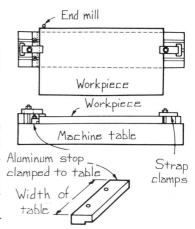

2.36 Milling-table T-square speeds 90° setups

Here's an easily made fixturing component that really simplifies square setups on a milling machine table. Dimensions can be adapted for use on any T-slotted table. Just take a couple of short lengths of 3/4-in. or 1-in. square bar and weld them at right angles with a triangular gusset of 1/4-in. plate as shown in the drawing. Then prepare the necessary holes for two press-fit 5/8-in. dowels or hardened pins and two T-slot bolts. The dowels, which key the fixture in the rear T-slot, should project out the bottom about 1/2 in. And finally, shim the fixture up off the table and mill along the working edge to make sure it's square.

GENE BRIGHTHAUPT, *Newark, Del*

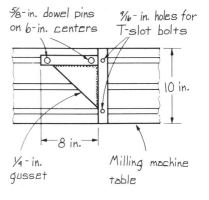

2.37 Mill-vise modified to prevent jaw lifting

We had a problem with the movable jaws of our universal milling vises spreading or lifting. This would cause parts to slip, and caused difficulty in holding flatness—a problem that was compounded when parts protruded above the jaws. As a result, many parts could not be machined in the vises, and more expensive fixturing was required. Trying to keep the play out of the movable jaws was a constant maintenance problem.

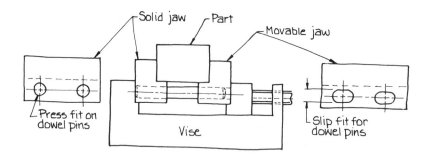

A relatively simple solution to the problem was the addition of two 5/8-in. press-fit dowel pins to the fixed jaws, with two matching slip-fit slots in the movable jaws, as shown in the sketch. The slots allow for minor clamping irregularities on the part clamping surface.

This fix virtually eliminated all play in the movable jaws, even on worn vises. It allows us to run many larger parts in vises rather than in more-expensive fixtures. And it has reduced our vise maintenance to a minimum.

DON J SCOTT, *Bellefontaine, Ohio*

2.38 Mount mill vise off center

In most job shops, it seems, milling machine vises are put on or removed from the machine tables every couple of days. And most of the time, they're mounted right in the middle of the table.

A better idea is to mount the vise about 8 in. off center.

This accomplishes two things: First, it provides more room on the table for setups that don't use the vise, so it doesn't have to be removed as often. This can save a lot of shop time normally spent in squaring up vises when they're replaced on the table. And second, it evens out the wear on the machine's table slides and feed screws—spreading out the wear to the ends, rather than concentrating it in the center.

GARY KIPP, *Wickliffe, Ohio*

2.39 Multipurpose bars have variety of uses

A set of several parallel blocks with the dimensions shown on the sketch can be extremely useful for tooling the table of a Bridgeport milling machine. The sketch also shows two ways they can be used as clamps, but that's only the beginning.

The 5/8-in. thickness allows the blocks

to be tapped into the T-slots of the table to square up work or position it for clamping. The 5-in. spacing of the outer holes equals the spacing of the outer T-slots, so the bars can also be bolted down as stops across the table, or they can be used as straddle clamps to hold relatively narrow work. And the central hole also allows them to be used as ordinary strap clamps.

Further, the 1 7/8-in. width of the bars suits them ideally for use as parallels in a Bridgeport milling machine vise, which has a jaw depth of 2 1/8 in.

The versatility of the bars will be further enhanced by making the holes oval.

WAYNE GOODRICH, *Springvale, Me*

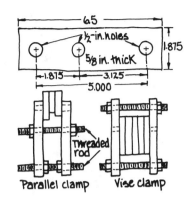

Parallel clamp   Vise clamp

2.40 Parallelogram T-nuts

As a production engineer, I was given the task of increasing the productivity of our plant's floor-type horizontal boring machines, the hourly rates for which are quite high.

I noted that every time a job was removed, the floor-plate T-slots were filled with chips, dirt, and muck of all sorts. A typical worker would require anywhere from four to eight hours to clean out the floor-plate T-slots to remove the T-nuts and reposition them for the next job—a very costly four to eight hours while the machine was not in use.

I made up some rhomboidal (parallelogram) T-nuts, as shown in the sketch, and flooded the boring department with them. They're simple and easy to insert—without having to clean out the full length of the T-slot. They virtually drop into place, and they engage by a simple clockwise rotation, which tends to tighten as the bolts are turned down. This not only increased the productivity of the boring machines, but operators

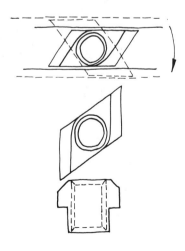

of all other machines have demanded the rhomboidal T-nuts as well.

C R NANDA, *Bombay, India*

2.41 Plastic quill stop

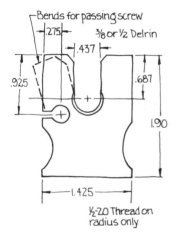

Several years ago, the AMERICAN MACHINIST published a Practical Idea showing a quill-stop for Bridgeport milling machines that was fly-cut so that it would pass the screw. The sketch here shows an improvement on that, which is made of Delrin plastic sheet 3/8 in. or 1/2 in. thick. Advantage of using plastic is that its flexibility allows it to be simply snapped over the screw, which, in turn, makes it much easier to make. And it's very handy.

ROBERT E LEUTZ, *Redwood City, Calif*

2.42 Plug is 'key' for safe start

We recently ran a job that required some internal machining in a reamed hole that

was first used to locate the part in the fixture. The location function was simply and accurately achieved by means of a plug through the part and into the fixture. Obviously the plug had to be removed before starting the cycle on the machining center being used, so that the tool could enter the hole to do its work.

To guarantee that this plug had been removed, we turned its unused end down to 3/8 in. diameter, and we covered the "Cycle Start" button with a plate having a slightly larger hole. This prevents the operator from cycling the machine with his finger, and makes a "key" out of the locating plug so that it *must* be removed from the fixture before the machine can be started (see first photo).

When this job is not being run, we simply swing the cover plate out of the way (second photo) so the operator can use his finger normally.

The operator tells us that the cover has prevented a crash more than once when, in the rush of production, he has forgotten to remove the plug and tried to cycle the machine with his finger.

P Aaron Lesko, *Allentown, Pa*

2.43 Precision fences speed setups on mill table

Here's an idea we've used for some time and which has saved us considerable time and money. The drawing shows precision fences that are self-keyed to a milling-machine table to facilitate repetitive drilling or milling operations without any need for truing up each piece with an indicator. They are simply slid into place and clamped with standard T-slot bolts. An edge finder is then used to establish the zero in both X and Y.

The fences are made of heavy angle or flat-ground stock and are milled on the bottom with projections that fit the table slots. Then they are drilled for the

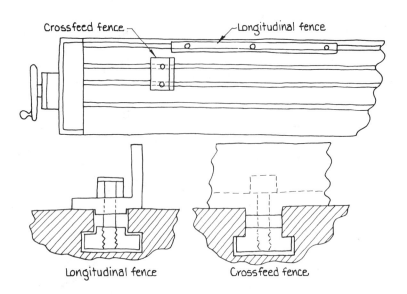

hold-down bolts. When first installed, it's a good idea to take a skim cut along each vertical face to ensure parallelism with each axis.

JAY HARFORD, *Annandale, Va*

2.44 Rolling tool slices stainless circles

The setup illustrated was used to cut a large order of stainless-steel disks 1/16 in. thick and 4 1/2 in. in diameter with 3/8-in. drilled holes in the center. The tooling, which was used on a Bridgeport milling machine, consists of an adjustable flybar mounting the roller from a pipe cutter, a backup plate fixed to the machine table, and a large hold-down disk.

LEE C WILKERSON, *Keysport, Ill*

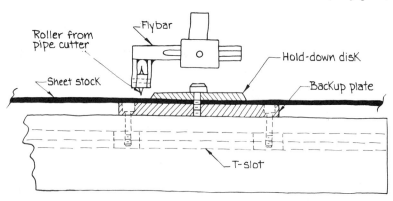

2.45 Set drill with planer gage

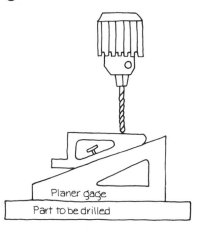

Planer gage
Part to be drilled

Whenever we ran our NC milling machine on jobs requiring many different-sized drills, it was a problem because of the many different collets or drill chucks that might be needed. Now we run most jobs with just one drill chuck to hold many different drill sizes.

The first drill is set to drill through the plate, and we then set an adjustable planer gage as shown in the drawing for setting all subsequent drills with the Z-axis retracted. Obviously, each tape must include a tool-change stop every time a new drill size is required.

Also obvious is that all drills should be approximately the same in overall length. But if you must go to a series of drills that is, say, 1 in. longer, just program the Z-axis for a 1-in. difference in work height. Then you can still use the same planer-gage setting.

RICHARD JORDAN, *Wickliffe, Ohio*

2.46 Setting up a swivel vise

To simplify and speed the setup of a vise that swivels around a center pin, I use the following procedure:

With a dial indicator mounted on the head of the mill, take a first reading directly in line with the pivot pin—approximately at the center of the stationary jaw. Then move to the end of the jaw and swivel the vise to repeat the original indicator reading. A slight touch up of position may be needed, and a final check should be made, but many passes across the full length of the jaw will be eliminated to save time and a lot of hand-cranking.

JOHN HUTTEMEIER, *Armstrong, Iowa*

2.47 Setup for angled operations

We had some shafts to make that featured milled flats at 60° to a drilled hole. To

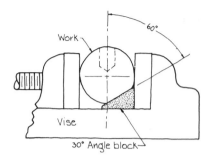

ensure consistent angular positioning while keeping fixturing to a minimum, I simply made a 30° angled block that fits in the milling-machine vise, and, as I tighten the vise, I lightly tap down the shaft with a lead hammer to ensure that each shaft is sitting firmly on the angled block. Note that the high side of the block goes against the fixed jaw of the vise to guarantee dimensional consistency and repeatability.

JOHN R MAKI, *Beverly, Mass*

2.48 Simple holder for boring

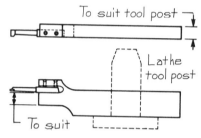

When boring small or precise holes in a lathe, it is always a hassle to use conventional boring equipment. I made this holder to take a standard toolbit blank ground to whatever size or shape is required when the holder is held in the toolpost without its rocker. In this way, the boring tool is always level with the bedways.

The tool slot in the holder can be milled to suit the largest toolbit you're apt to use, and any smaller size can be used without any problems. The holder I made is ordinary cold-rolled steel, but, although it works fine, use of properly heat-treated tool steel would extend the life measurably.

HAROLD M SHAFLEY SR, *Mineral Wells, Texas*

2.49 Simple T-slot stops

These double-wedge T-slot stops hold firmly in position and at the same time accommodate themselves to slight variations in T-slot width.

To make the stops, I begin with some 5/8-in. low-carbon stock and mill the wedges in pairs to make certain that the angles are identical on mating pieces and thus the sides are parallel when assembled. Then each set is held vertically in a vise and adjusted so that the overall width of the pair matches the T-slot width. In this position they are tap-drilled and then clearance-drilled to within 0.150 in. of the bottom. Finally, the bottom piece is tapped.

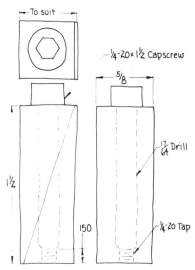

Two points on the above: The clearance drill should be generous, a 17/64 in. or even a 9/32 in. diameter. And there should only be enough stock left at the bottom for two or three threads. Both of these factors allow for some misalignment of the axial capscrew when the wedges are drawn together.

In use, the screw is backed off a bit, the stop is inserted in the T-slot, and the screw is tightened. Hand tightening is usually sufficient, but if a very firm stop is desired a light torquing with an Allen wrench is all that is needed. The wedge angle is steep enough so that the stop is self-releasing when the screw is again loosened.

RAYMOND B HARLAN, *Wayland, Mass*

2.50 Slot broached last in NC program

The casting shown in the photograph is completely machined in a two-station fixture on the table of a tape-controlled Model 4JE Giddings &Lewis horizontal boring machine. The only remaining operation is broaching a 1/2-in.-wide slot in the ID of a 2.26-in. hole. We decided to try broaching on the NC machine because the part is very difficult to load into a broaching machine and it's an extra operation in any case.

The broach is held in a Davis holder (306-10292) with a 0.500-in.-dia pilot hole; it is not locked, but floats—stopping against a shoulder on the broach. A broach horn is inserted in the hole and secured against rotation with a 1/4-in. pin through the horn flange into a mating hole drilled in the part.

Except for manually moving the W-axis to engage the broach pilot after inserting the broach into the horn, the operation is done on tape in four passes.

After the first pass, an 0.063-in. shim is put under the broach, and the machine positions to a new location, keeping the broach level.

To enable the machine mode, the spindle is programmed for 6.9 rpm, but the spindle is put in neutral and locked to prevent rotation. The Z-axis is set to a predetermined position and only the W-axis is used for motion. A feedrate of 49.4 ipm has proved to be successful in our application.

RICHARD H ROWE, *Loves Park, Ill*

2.51 Spacers for depth adjustment on a mill

On a vertical milling machine, when you feed the quill for a short distance and want a positive stop, it becomes a nuisance to run the micrometer adjusting nuts up and down the quill-stop screw. I've made myself a set of removable spacer sleeves that fit on top of the adjusting nuts, which eliminate the need to position the nuts more than an inch.

The sleeves can be made of 1-in. or 1 1/4-in. barstock bored out and then slotted on one side to clear the screw and the hub on the top adjusting nut. The set consists of sleeves 1, 2, 3, and 4 in. long, which were hardened and ground to produce a permanent, time-saving attachment.

GENE BRIGHTHAUPT, *Newark, Del*

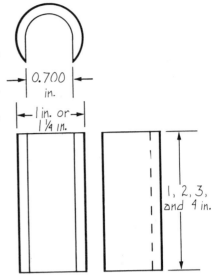

2.52 Special tooling for oversize backfacing jobs

In big fabricated structures or heavy frame castings, machining of inner flange

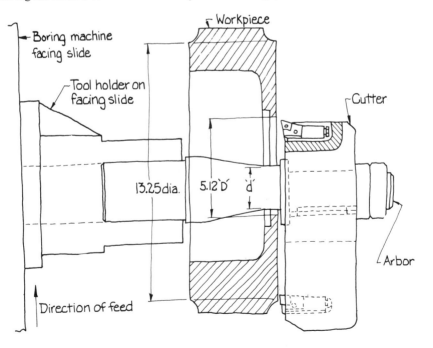

faces often becomes quite difficult because typical back-facing cutters are limited in the diameters they can machine. The tool shown in the sketch was designed to machine an inner flange of 13.25-in. diameter with a central access hole of only 5.12-in. diameter. It was used on a horizontal boring machine equipped with a facing slide. Details are shown on the drawing.

In use, the arbor is first inserted through the bore and the cutter body is mounted and positioned as shown. When the faceplate slide feed is engaged, one of the insert cartridges moves outward and the other moves inward. The basic cutter design permits the machining of a maximum flange diameter of $4(D - d) + D$, in which D is the central bore diameter and d is the arbor diameter where it passes through the bore.

Y KUMARAPPAN, Trivellore, India

2.53 Special T-slot clamp is faster than strap

The drawing illustrates a type of special T-slot clamp that we find is quicker to use for fixturing jobs on our Bridgeport milling machines than conventional clamping systems. They slide in from the end of the table: just angle the clamp, insert the front lugs, slide the clamp in a few inches, and then drop the rear lugs down into the coolant recess for insertion into the T-slot and then slide the clamp up to the workpiece. It's quicker to do than to describe. Also, to avoid marring any previously machined surface, put a pad (cut from 1/4-in. or 1/2-in. plate) under the clamp screw.

BASIL CORLEY, Louisville, Ohio

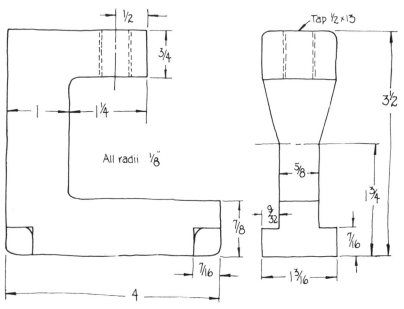

2.54 Split spacer for milling arbor

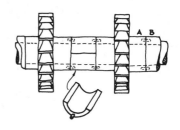

We received a quantity of castings that required a straddle-milling operation. It was quite straightforward except that one section was wider than the other. We didn't want to run each casting through twice, and we felt it would be too time-consuming to pull off the spacers and one cutter and reassemble them on the arbor.

To speed the change, we made some special arbor spacers. One was split, and each half has two short dowels, as shown. We then made four additional spacers with mating dowel pin holes.

The first milling cut is made with the setup as shown. The arbor nut is then loosened a few turns, the split spacer is removed and replaced outboard between collars A and B (which are drilled for the dowel pins that stay with the split arbor halves).

CLINT MCLAUGHLIN, *Jamaica, NY*

2.55 Spring holds quill clamp on mill

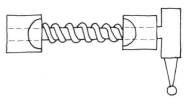

Vibration tends to tighten up the quill lock on my Bridgeport. Most machinists use a dollop of clay or a rubber band to hold up the locking lever, but I fashioned a more permanent fix.

I disassembled the lock and inserted a mild compression spring between the two bronze gibs. This spring is a loose fit over the bolt, but it provides enough force to hold the gibs apart, or at any particular tension setting on the gibs.

DAVID R CARLSON, *Manchester, NH*

2.56 Spur gear is indexer

Milling of keyways (or slots for Woodruff keys) at a specific offset angle on a shaft can easily be done by using a spur gear as an index plate. It may be

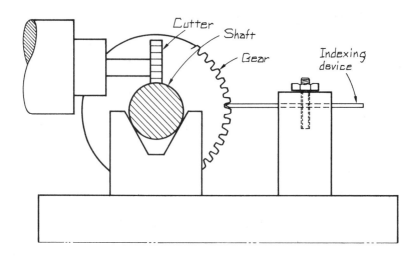

necessary to bore out the gear to fit the shaft, and the hub may require drilling and tapping for a setscrew to hold it on the work.

With the gear in position, the workpiece is set in V-blocks or a fixture, the first operation is performed, the work is indexed (using the index finger in the gear teeth), and the next operation is performed. To calculate the number of teeth the part should be rotated, just divide 360° by the number of teeth on the gear. A 72-tooth gear will index 5° per tooth, a 36-tooth gear will index 10° per tooth, etc.

Jim Willett, Springville, NY

2.57 Step-block shims four different dimensions

It's often necessary to accurately mill finished dimensions on four sides of square or rectangular blocks that have been rough-sawed or flame-cut to slightly oversize dimensions.

When the job entails making a number of such pieces, a very simple stepped block, slightly narrower than the part thickness, can speed the job by shimming up four parts to each respective new dimension, as shown in the sketch. The height of the steps must be calculated to arrive at the finished dimensions.

The step-block is placed in the bottom of the milling-machine vise, and all four parts can be clamped securely by using a strip of leather or plastic against the movable jaw.

After each complete pass of the milling cutter, each part is rotated and advanced to the next step, with a new block placed on the first step. Then each pass of the cutter produces one completely finished part. Obviously, you start with one piece, then two, and so forth.

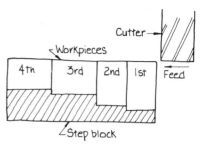

HERBERT FRANK, *Philadelphia, Pa*

2.58 Stepdrills stretch ATC

We machine a large variety of aluminum castings on CNC machining centers—and we often face a problem of limited tool-changer capacity that prevents doing a job complete in one operation. In programming one of these workpieces recently, for example, we came up short by just one position in the tool-storage drum of the automatic toolchanger.

We were able to overcome the problem, however, by using a custom-made stepdrill. The part required a number of 1/2-in. and 3/4-in. pipe-threaded holes. By grinding a stepdrill with 23/32-in. and 59/64-in. diameters, we were able to use a single tool—and only one position in the toolchanger—to tap-drill both hole sizes merely by programming the drill to different depths, as appropriate.

Ever since that time, we have used stepdrills for a variety of applications in which part configurations permitted this approach. In general, castings with a number of cavities seem to offer the most opportunities for taking advantage of the stepdrill technique.

JOSEPH D JUHASZ, *Elkhart, Ind*

2.59 Subplate is first job to run on NC machine

A subplate is an absolute must for any NC machine; it should be the first job that is run on it. The subplate will speed making setups, and will go far to eliminate misalignments of fixtures and indexers. On smaller machines, it can also be used to increase the usable machining area.

Use a piece of 1-in. to 2-in. steel plate that is large enough to cover the entire machine table. For large machines, this can be done in two pieces, which makes it easier to work with. And if the table is small and you plan on using an indexer on it, use a piece long enough to cover the table plus an extension area to allow mounting the indexer off to one side of the machining area.

Use the machine's maximum X and Y travels to determine the maximum-size hole pattern and center it on the subplate. Measure the T-slot spacing and put

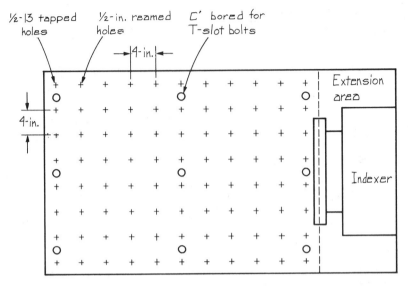

in counterbored holes for hold-down bolts, and then use them to mount the subplate on the machine table with some blocks underneath for clearance above the table. Now machine the hole pattern—a 4-in. pattern alternating tapped and reamed (or bored) holes is convenient for our work—in either manual or automatic mode. Now turn the plate over, deburr all holes, remove the riser blocks, clean mating surfaces, reinvert the plate, and bolt it loosely in position. Using two of the end dowel holes, indicate the plate parallel with the X-axis, and tighten the hold-down bolts.

All fixtures, indexers, and other workholding devices should have at least two dowel holes in their baseplates for alignment with the dowel holes in the subplate, and if clearance holes can be provided for bolting to the subplate it will eliminate the need for strap clamps. By indicating a dowel hole in the subplate, you can determine the exact location of all dowel holes and use them as an aid in programming and fixture design.

JOSEPH D JUHASZ, *Horsham, Pa*

2.60 Three pistons ganged for heavy clamping

The task was to hold a heavy workpiece for a production milling operation.

The problem was that insufficient space prevented use of a heavy-duty hydraulic cylinder.

The solution was a specially designed cylinder assembly made in our shop. As shown in the sketch, the design puts three pistons in a tug-of-war series that multiplies clamping force. Pneumatic pressure is applied to each of the cylinders in the assembly by way of an exterior valve, and interconnecting passages equalize the pressure in each one.

Milling and Boring

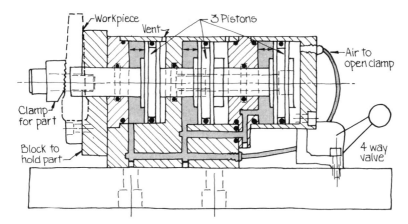

The result is a tripled clamping force that holds the work securely throughout the milling operation.

GEORGE BO-LINN, *Houston, Texas*

2.61 Three-bit flycutter gives better finish faster

I have been able to increase feedrates 100% and still obtain a better surface finish with the three-bit flycutter shown in the drawing, which gives dimensions for both 3-in. and

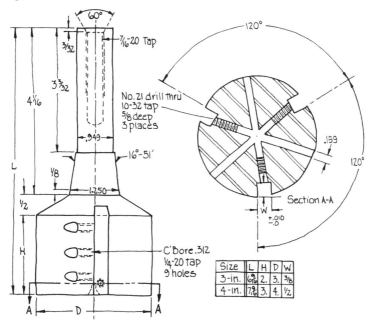

Size	L	H	D	W
3-in.	6⁹⁄₁₆	2.	3.	³⁄₈
4-in.	7⁹⁄₁₆	3.	4.	½

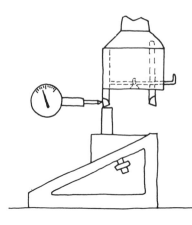

4-in. versions. Longer tool life and reduced vibration are also evident.

Lathe toolbits (I use the brazed-carbide type) are adjusted to eliminate radial runout with a dial indicator, and axial runout of the cutting edges is adjusted with a solid stop, such as a vise jaw, planer gage, etc., as shown in the smaller sketch. Once the toolbits have been tightened securely, the flycutter can be used and removed as needed.

An R-8 shank is shown on the drawing, but any others may also be used, depending on the machine to be used.

WILLFRED G MOORE, *Chicago, Ill*

2.62 Tool-stand holds milling-machine shanks

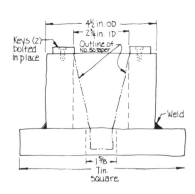

The bench stand sketched here greatly facilitates such toolroom chores as indexing the inserts in face mills, changing drills, and similar tasks. Although the unit shown is designed for use with No. 50 tapered toolholders, smaller versions could as easily be made for No. 30 or No. 40 tapers, or, indeed, for any style toolholder.

The stand shown consists of a 7-in. by 7-in. base made of 1-in. hot-rolled steel plate with a 1 5/8-in. hole drilled in the center, to which a 4-in.-high piece of tubular stock (2 3/4-in. ID) has been welded. A pair of keys (each 1 in. x 1 in. x 1/2 in.) is bolted on top to take the torque of tightening and loosening the tools themselves.

ED KOELLER, *Holland, Mich*

2.63 Trig helps tool miller table

Producing angled setups on a milling-machine table is often easier with a pencil and paper—or a hand calculator—than it is with a protractor.

Horizontal angles are easily set as shown at 1 in the drawing. Set two snug-fitting dowel pins any distance (A) apart. Then, to offset the work by some specific angle, look up the sine of the angle and multiple by A to find the width of the block needed. Making A equal to 10 in. is especially convenient.

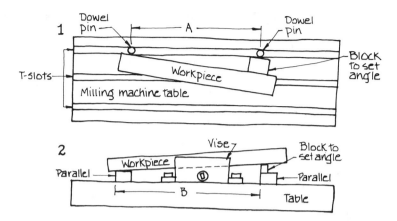

The same basic idea can be used for setting work at specific vertical angles in a mill vise. In the setup shown at 2, however, B should be multiplied by the tangent of the desired angle to find the thickness of the setup block.

THOMAS M TOMALAVAGE, *Roselle Park, NJ*

2.64 T-slot stops feature built-in clamps

Just take some rectangular bar stock of suitable dimensions, cut it into the desired lengths, and drill and tap a pair of holes in each length (as shown in the accompanying sketch), add two short hex-head machine screws to each, and you have a set of parallels or end stops that clamp conveniently into the T-slots of your milling-machine table. Mine are made to the dimensions indicated, but they could be made to other dimensions, as well. Just be sure that the stock is not too thick to prevent an open-end wrench from being inserted into the T-slots for clamping.

JEFF FRECHETTE, *East Lyme, Conn*

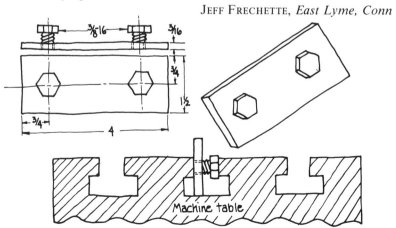

2.65 Vise jaws reduce need for special fixturing

The sketch illustrates a set of universal vise jaws which we use for workholding on our Bridgeport BTC II NC machining center. Having installed a pair of Kurt vises on the machining center, I was confronted with the problem of making universal jaws to eliminate the costly and time-consuming requirement of making special jaws for different workpieces.

The biggest problem encountered turned out to be clearance under the part for tooling. This was solved in the illustrated design by using removable pins, which can be placed in any holes suitable to provide tool clearance for the particular workpiece involved.

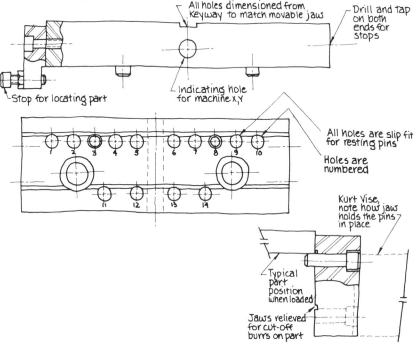

The drawing shows only the stationary jaw; the movable jaw is simply a mirror image, but with some of the details omitted, such as the indicating hole and provision for end stops which should not be on a movable jaw for reasons of precision.

Note also on the drawing that all holes are individually numbered. By recording which holes are used in setting up for any particular workpiece, subsequent setups for the same part can be speeded.

The original jaws are now two years old and have been used to produce more than 50 different parts to date.

JIM KRAFT, *Florissant, Mo*

3 Turning

3.01 An accurate setup for turning perfect tapers

Here's a procedure for setting the compound feed of a lathe to an exact angle for accurate turning of tapers:

Mount an angle plate on the lathe faceplate or the face of the chuck and rotate this so that its outer face is perpendicular to the lathe bed.

Position the compound to the approximate angle, snug up the bolts, and adjust the compound as far backward as needed to provide the forward travel, take out the screw backlash, and set the compound dial on zero.

Move the carriage and cross-slide into position, and pick up zero on the dial indicator mounted in the toolpost.

Look up the sine of the desired angle (0.1305 for the 7° 30' of the example), and multiply this by the feed distance of the compound. (I find 2 in. or 3 in. quite suitable.) Using the compound dial, move the compound forward by the distance selected.

Adjust the angle of the lathe compound as shown in the drawing, using gage blocks totaling the thickness calculated above, until the dial indicator reads zero at both positions shown. Then clamp the compound.

FORREST N MOHLER, *Nineveh, Ind*

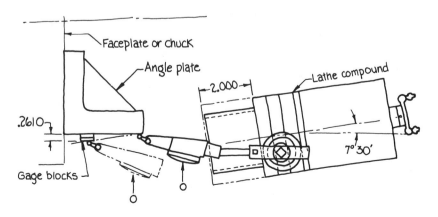

3.02 Add form-tool adjustability

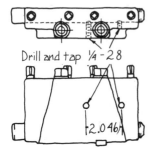

In our shop we must hold bevel-gear angles to ±0° 4'. The angle is finished as the part comes off a six-spindle automatic, which necessitates a very accurate means for adjusting the form-tool used to produce the angle.

This was done very simply and effectively by drilling and tapping a pair of 1/4-28 holes in the side of the form-tool holder spaced 2.046 in. apart. We then formed spherical radii on the ends of two 1/4-28 setscrews, so that the tool shank could "rock" around these ends.

One complete turn of either screw produces an angular change of exactly 1°, and a fraction of a turn changes the setting by the same fraction of 1°.

T S COTTER, *Perry, Ohio*

3.03 Add mechanical tracer to toolroom lathe

Practically any toolroom lathe can be turned into a tracer lathe quickly and cheaply. Rigidly secure a flat steel bar or plate across the lathe bed near the headstock on which to mount shaped templets, which may be radiused, angled, or otherwise contoured. Drill and tap two holes in the edge of the cross-slide and attach a T-shaped roller holder so that the roller (a suitably sized ball-bearing) can be snugged up against the templet. Now add a couple of die springs between the saddle and the tailstock with enough preload to keep the roller forced against the templet.

With the saddle unclamped and the leadscrew disengaged, the saddle will now follow the contour of the templet when the cross-slide feed is engaged.

The position of the roller and templet in relation to the workpiece is not important, but the position of the toolbit is critical. To set up for a concave job such as the one shown in the larger drawing, crank the cross-slide slowly back and forth with an indicator on the templet to determine the low point. Then, without moving the cross-slide, center the toolbit on the work both horizontally and vertically. To produce an annular groove, as shown in the small sketch, offset the toolbit by an amount X that is equal to the desired radius of the annulus.

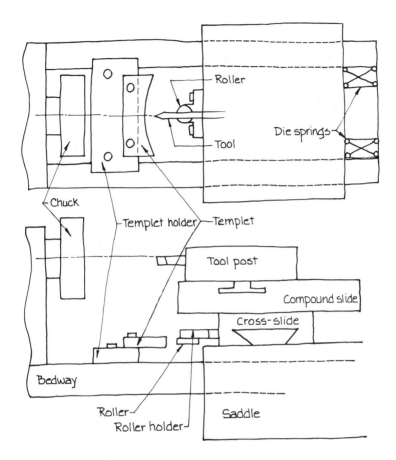

No sizes or dimensions have been given because the method can be used on many different lathes and it is easy to adapt the "attachment" to suit.

ANTHONY G CARR, *Wakefield, Mass*

3.04 Adjustable collet stop

The collet stop shown in the drawing is simple to make, inexpensive, and easy to adjust to whatever length is required. I've been using one like it for years on Hardinge toolroom lathes, and it's very positive in repeating lengths even though the diameter of the workpieces may vary. Another advantage is that its length setting can be adjusted without removing the collet.

Basically it's nothing more than a 1-in.-dia aluminum bar with a long flat at the rear end to prevent the setscrew from raising any burrs that would interfere with axial settings and with the front end bored out to accept adapters for

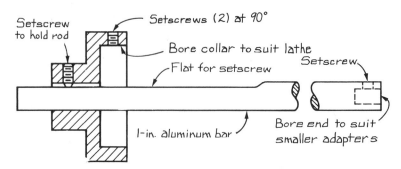

smaller diameter jobs on which the stop must project into the rear of the collet. This bar is held in a simple collar that is secured to the lathe spindle by two setscrews spaced 90° apart.

RICHARD ROLSTON, *Simi Valley, Calif*

3.05 Adjustable spacers gang up for bigger jobs

Here's an idea for adjustable spacers to be used in conjunction with micrometer stops or fixed stops in a lathe. I use 1/4-20 hex-head bolts with jam nuts in the ends, setting absolute length with either a micrometer or a height gage. I've made a set of the stops out of 1/2-in. cold-rolled steel with lengths of 1 in. through 6 in. in 1-in. increments. For spacers over 6 in. long, just put two or more together with headless 1/4-20 setscrews.

ED ZATLER, *Westland, Mich*

3.06 Another chuckable center

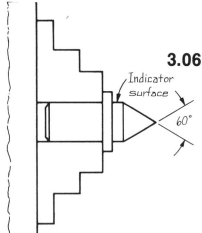

Holding a lathe center in a chuck is a quick way to switch from chucking jobs to between-centers turning, and the 60° point can be trued easily for accurate work (if the center is not hardened).

The lathe center illustrated also prevents it from being forced back into the chuck jaws by tailstock pressure. And if the short cylindrical section just behind the conical point is turned at the same time, it can be indicated to true it accurately if it is removed and then re-chucked. In addition, the indicator surface can be useful in setting up for an eccentric turning job.

WILLIAM D ROBERTS, *Portage, Mich*

3.07 Attachment speeds drilling from tailstock

The drawing illustrates some tooling we made up to speed the job of drilling in our engine lathes.

The unit consists of a Morse-taper sleeve tapered to fit the tailstock and bored and reamed to provide a straight, concentric central hole. Toolholder bars (three typical ones are drawn) are made up to slide in this central hole, and the business end is configured to hold whatever type drill or chuck you plan to use. Just behind this is a cross-hole for feeding the tool.

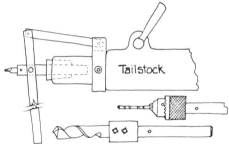

Clamped to the tailstock quill is a steel collar on the rear side of which are welded two lugs to support the feed lever system. The front end of the main lever consists of two bars in a clevis arrangement, one above the toolholder bar and the other below, which is drilled vertically to match the cross-holes of the toolholder bars. With a toolholder bar in place, its cross-hole vertical, a headed pin is dropped through the aligned holes, and the attachment is ready for use.

CLINT MCLAUGHLIN, *Jamaica, NY*

3.08 Backup tube reinforces pipe-turning spider

Use of a backup tube to reinforce a lathe spider, as shown in the drawing, will enhance both the safety of the setup and the accuracy of the machining. The internal backup effectively prevents the possibility that the spider will "walk" down the inside of the workpiece, and it thereby reduces the necessary outward pressure of the spider's "legs," which will reduce any distortion of the work.

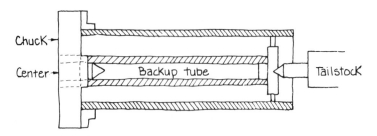

The spider itself should be machined with a "spigot" on its rear surface to support one end of the backup tube, while the other end should be supported by a fixed center in the spindle.

JOHN L CROMPTON, *Palo Alto, Calif*

3.09 Ball-turning attachment fits on lathe turret

Here's a turret-lathe attachment we made for turning a ball on the end of a length of barstock. The device consists of a block bored out to accept a shaft with a worm-wheel mounted on the rear end and an offset bar at the front to carry the cutting tool. The assembly is mounted on the lathe turret with a tapered shim block to set it at the proper angle, as shown. Two plates are mounted above and below the bored block to support a short, vertical shaft with a worm-gear to engage the worm-wheel. A crank handle is fixed at the

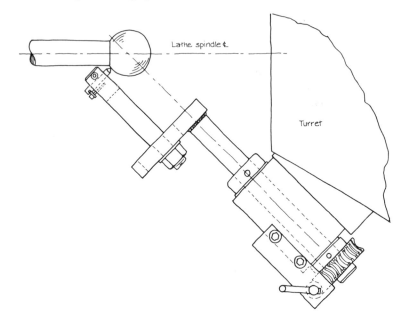

top of this vertical shaft for rotating the worm. As this crank is rotated, the cutter swings around the ball, cutting it to shape.

The end of the toolholder bar has a square hole for the toolbit, and the end is split to allow a capscrew to clamp the bit securely. The toolbit is fed into the work by means of a small socket-head screw with a washer brazed under the head to engage a slot in the bit. These two screws are used to unclamp, advance, then reclamp the bit until final size is reached. There is a small area at the end of the workpiece that must be finished with a file or a form-tool.

CLINT MCLAUGHLIN, Jamaica, NY

3.10 Bar-puller fits turrets of NC lathes

Our company recently purchased a pair of Okuma Model LC-40 CNC lathes, which have proved to be both highly productive and easy to program. At the time we ordered them, we also considered the purchase of bar-feed units. Because their installation would have required a considerable amount of floor space and because of the expense involved, the company decided against acquiring these accessories.

Thus, to automate the feeding of stock in bar operations, we really had to come up with a simple and inexpensive alternate method.

The opportunity for this was provided by the fact that these machines have dual independent turrets that can be operated simultaneously. What was devised was the simple mechanism shown in the sketches, which I call a "bar-puller." Two identical units are made and the pair are mounted in the dual turrets opposing one another, as sketched at the right.

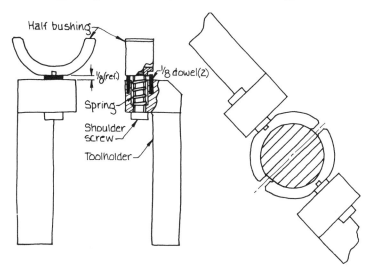

The shoulder screw shown in the drawing is a 5/16-18 with a 1 1/4-in. shoulder length; the springs are 5/8 in. OD and rather heavy-duty for a firm grip; and there's 1/8-in. "give" in the grippers.

The two turrets are then programmed to close in on the stock simultaneously and then—still simultaneously—to move the desired stock-feed distance in the Z-axis. A length of barstock is loaded, and the program is run in a loop to relieve the operator of unnecessary chores. In our case, this permits one operator to run both machines.

JOSEPH GOLDENBERG, Brooklyn, NY

3.11 Bar-puller positions work

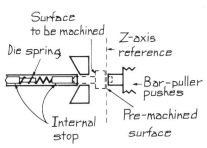

Maintaining a required tolerance to a premachined surface opposite the internal chucking stop can sometimes be a problem on an NC turret lathe. Refinishing the machined surface in the same chucking as the new cut might be one answer. But perhaps better is a method possible on machines with bar-pulling capability, such as Warner & Swasey SC machines.

Add a die spring to the internal stop so that it offers resistance to the part as the part is pushed by the bar-puller. After the part is chucked, the puller is programmed to push the part to a reference position, after which the program calls for clamping the part. The program can now index the turret to the desired cutting station for machining the new feature to the premachined reference surface.

PAUL F SUTPHEN, Power Controls Div,
Midland-Ross Corp, Owosso, Mich

3.12 Barstock soft jaws grip different diameters

Too many times when a set of soft jaws are needed, they just aren't available. Finding myself in that position, I cut three 1-in.-thick disks from some 3-in.-dia aluminum barstock, facing both flat surfaces and drilling and counterboring a clearance hole in the center of each for a 3/8-in. socket-head screw. After removing the top jaws, I mounted the aluminum disks to each of the

Turning

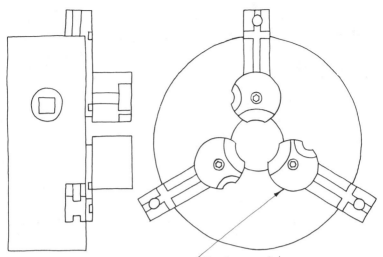

Aluminum soft jaws

chuck jaws with a single jaw bolt for each. Finally, the aluminum barstock soft jaws were bored out to receive the work they were made for.

These barstock soft jaws could as easily be made of steel, but I prefer aluminum because they will not mark the surface of parts being machined.

After the job is finished, the soft jaws can be put away for future use, at which time they can be rotated to an unused area and rebored to whatever diameter is then required.

JOHN URAM, *Cohoes, NY*

3.13 Bench holder for 5C collets

Here's a time-saving fixture for securely holding 5C collets for setup with threaded stops. The fixture is machined to fit the rear body diameter of the collet and the collet keyway; other dimensions are completely uncritical. I hold the fixture in a vise, but it could easily be bolted to a convenient bench or even the side of the machine for quicker access.

GENE BRIGHTHAUPT, *Newark, Del*

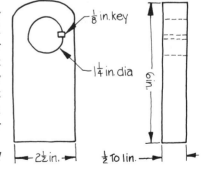

3.14 Boring-bar support in lathe ends chatter

Most machinists have encountered chatter when boring holes of long depth/diameter ratios. The sketch illustrates a setup that overcomes the problem.

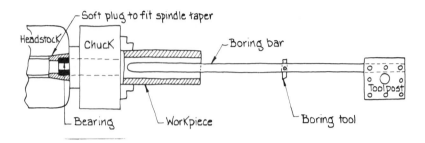

Turn a piece of mild steel or aluminum to fit the spindle-nose taper, set it in the spindle with a mallet, and then drill and bore the plug to accept a bearing blank (nylon or Delrin are good for vibration damping). The bearing is then bored for about 0.001–0.002 in. clearance over the boring-bar diameter.

The boring bar is a length of drill rod, straight and free of nicks or burrs. Taper or radius the front end for a lead-in into the bearing, and center-drill the rear end. For setup, suspend the bar between the spindle bearing and the tailstock center and clamp it in the toolpost with no deflection. Naturally, the cross-slide is not moved after this. And check that the bar moves freely through the bearing for the full travel necessary for the boring job. Depth of cut is set by tapping the toolbit forward and checking its position with a dial indicator.

This technique permits the use of higher spindle speeds and feedrates and significantly increased metal-removal rates. And finally, it also improves the parallelism of the bore.

ANTHONY G CARR, *Wakefield, Mass*

3.15 Brass face for chuck jaws

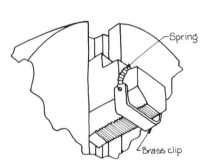

When it's important not to mar the finish on parts being chucked in a lathe, brass shims are often used to protect the surface. But it can be a struggle to insert a shim under each jaw, support the workpiece, and tighten the chuck all at the same time.

These easily made brass U-clips solve the problem. They're held in place on the chuck jaws by light coil springs, such as short lengths of garter spring, which provide sufficient expansion for fitting the clips over a wide range of different chuck jaws.

KEON GERROW, *Racine, Wis*

3.16 Broken saw makes cutoff blade

A broken slitting saw can be given new life as a lathe parting-tool blade simply by slicing out a suitable section as shown in the drawing. The job is done using an abrasive cutoff wheel. Specific dimensions, of course, will vary with the specific toolholders used in the shop. And it's also possible to produce a number of cutoff blades from a single broken sawblade.

K S PATIL, *Maharashtra, India*

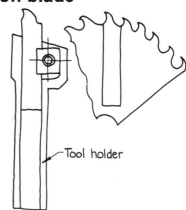

3.17 Bushing grips parts for eccentric turning

Sometimes we have to turn eccentric cams or cranks in our shop, and if more than one piece has to be made it becomes a repetitive nuisance to have to keep setting up in a four-jaw chuck with an indicator to determine the exact offset required.

It's a lot simpler and faster to do that just once to make a split bushing, as shown in the drawing, and use that as an insert in a standard collet to grip the workpieces off-center by the right amount every time.

BILL KAHN, *Mahopac, NY*

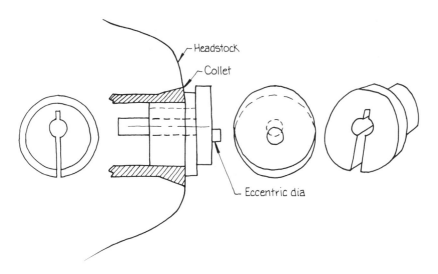

3.18 A calibrated cross-slide

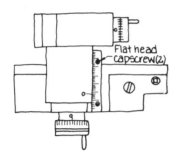

Keeping track of a lathe tool's position can sometimes be confusing when several turns of the cross-slide feedscrew are necessary between cuts. A digital readout provides the ultimate in convenience and accuracy. A far less costly, though admittedly rougher, solution is to fasten a 6-in. steel rule next to the cross-slide and stamp a witness mark at some convenient point on the cross-slide. I have found this very useful in eliminating full-revolution errors of the cross-feed dial.

DAVID C RUUSKA, *Maple Grove, Minn*

3.19 Cardboard protects chuck

Boring operations in a lathe are often performed with the chuck jaws opened wide enough so that the chuck threads or scroll are exposed and chips from the internal machining operation too easily get lodged within the chuck mechanism. This not only makes it difficult to open and reclose the jaws, but can result in damage to the chuck itself.

I avoid the problem by the simple expedient of making protector disks of heavy (approximately 1/16 in. thick) and stiff cardboard that fit snugly against the face of the chuck. To make a disk, I place the workpiece on the cardboard and trace its outline. Then I cut the disk about 1 1/2 in. larger in diameter and notch the periphery down to the part outline to clear the jaws.

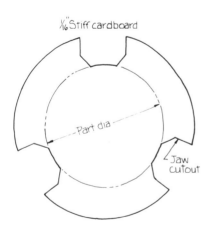

It works equally well with either three-jaw or four-jaw chucks, prolonging their lives and saving me considerable time in cleaning the chucks during or after use.

ROBERT ULMER, *Downers Grove, Ill*

3.20 Centerdrill holder fits lathe toolblock

This centerdrill holder, used with a block-type toolholder on the carriage and with the workpiece set to a stop, will ensure uniform and true center holes. The outside edge is set perpendicular to the chuck face with a try square.

When making the tool, set it up in the lathe's toolholder and centerdrill it from the chuck or collet to ensure that the holder's holes are properly on center and parallel to the turning axis. If several different size holes are drilled, a variety of centerdrill sizes can be accommodated.

The holder is very helpful in producing uniform center holes in workpieces, especially when the parts are to be ground on centers in a subsequent operation.

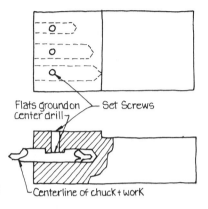

PHILLIP E ZATLER, *East Lansing, Mich*

3.21 Chucking thin disks

It is sometimes necessary—and usually difficult—to set a small disk-shaped part in a three-jaw chuck so that it runs true without wobbling. Here's an easy way to do it:

Apply a small magnet to one side of a 6-in. steel rule. Stick the workpiece onto the other side of the rule. The workpiece can then be readily set in the chuck exactly flush with the outer face of the jaws. And if the workpiece is nonmagnetic, use double-coated adhesive tape.

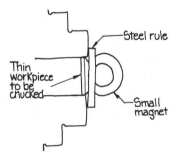

GUY LAUTARD, *W Vancouver, BC*

3.22 Collet bushing holds stubby parts

When the right-sized spring collet is not available for a given secondary operation, a two-piece split bushing cut from a single blank will serve the purpose. This is used in conjunction with an oversize spring collet.

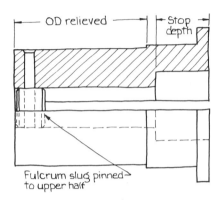
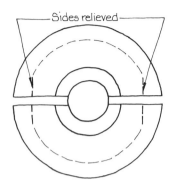

The design shown here not only overcomes the potential problem of insufficient gripping power on short, stubby parts, it also provides for its own axial positioning in the collet and for that of the workpiece as well.

The idea has been used in machining operations that require considerable cutting force, including the interrupted cutting encountered in polygonal turning with a special attachment. It's especially recommended for use in second-operation work on ram-type manual turret lathes and milling jobs in which collet fixtures are used.

S PADMANABHAN, *Madras, India*

3.23 Collets for toolholding

Doing a lot of varied turning jobs, I frequently have to change to different size boring bars—and it was always a bother to search for the proper holder sized for the shank of the tool I wanted to use.

But we had plenty of 5c collets, which would fit virtually any shank I might ever want to use. Realizing this, I made a fixture (see photo) that would hold a collet on the quick-change toolpost on any of our lathes. Basically, the fixture is made to fit the dovetail and it is taper-bored to fit a 5c collet. An adjusting nut on the back tightens the collet on the boring bar.

JOHN WILLIS, *Huntsville, Ala*

3.24 Cool the cut with air

When it's necessary to turn a fine finish to close tolerance on a lengthy portion of a shaft, I prefer to use a stream of air as the coolant.

The air nozzle is affixed to the toolpost about 1 1/2 in. from the tool tip and aimed at the forming chip. This provides better control of chip formation, keeps the work cool, and also keeps the tool and toolholder cool enough to prevent significant thermal expansion (which would reduce part diameter from the setting). In addition, the use of liquid coolant—whether flooded or applied otherwise—seems to deteriorate the carbide cutting edge too rapidly.

For the finish pass, a depth of cut of 1/16 in. to 1/8 in. produces a uniform, on-size finish—almost indefinitely.

DAN O'LEARY, *Calgary, Alta, Canada*

3.25 Corner-rounding end-mill as lathe tool

Sometimes the tedious and time-consuming task of grinding an inside radius on a lathe tool can be eliminated by substituting a corner-rounding end-mill as a radius-turning tool. A Brown & Sharpe 1 3/4-in.-sq chuck block employing 5-C collets can be used as a toolholder, and this can be clamped to the lathe compound with a short plate or strap-clamp with two holes drilled for T-slot bolts. A shim or spacer of the proper thickness will also be required to raise the chuck block so the end-mill cuts on center.

This method has proven very effective when standard-size radii are required.

GEORGE E HAYS, *Greensburg, Pa*

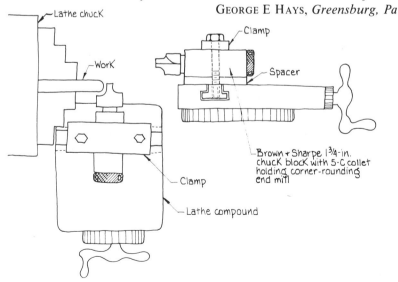

3.26 Cross-drilling turned parts in a hand screw machine

We recently turned a number of short, shouldered parts that also required drilling of a cross-hole. At one turret station of the hand screw machine, we mounted a sleeve with a cross hole to act as a drill guide. This sleeve should either be hardened or have a standard drill bushing pressed in.

An electric drill was mounted horizontally at the rear of the lever-operated cross-slide, as drawn. (Most supply houses stock standard bases for mounting electric drills to flat surfaces.) Because the drill tended to run hot if left running continuously, we added a flat-spring trigger actuator that was mounted on a bracket clamped to the machine bed.

To produce a part, the barstock is first turned. Then the spindle is stopped and the sleeve is fed over the part. Because the old machine did not have a spindle brake, a wooden wedge was pushed under the flat-belt headstock pulley. Next the cross-slide is pulled forward to drill the hole. And, finally, the wedge is removed, the spindle is restarted, and the cross-slide is fed back to cut off the finished part with the front tool.

CLINT MCLAUGHLIN, *Jamaica, NY*

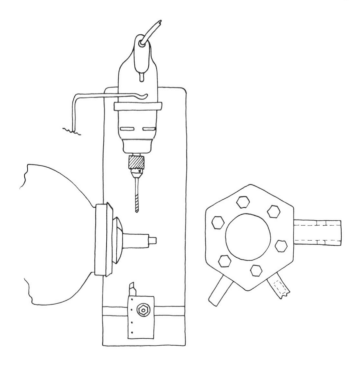

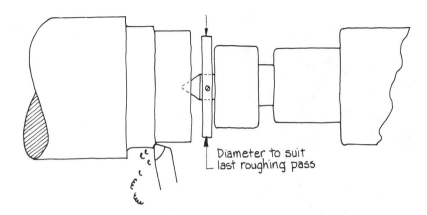

3.27 Disk on tailstock center alerts operator

As do most machine shops, we have a shortage of experienced machinists, and consequently our personnel includes many trainees. We give them as much responsibility as possible, but some of them work very slowly because they are afraid of scrapping a workpiece. The idea described here is aimed at increasing their confidence level when turning shafts in a lathe. It'll do the same for a journeyman.

When turning down the ends of shafts, we simply attach a disk to the tailstock center with a setscrew. Diameter of the disk is slightly larger than the finished diameter that's supposed to be turned. The operator just feeds the tool forward in line with the disk and need have no fear of going undersize as long as the disk is cleared. Once the tool gets down to the disk, the operator is warned to proceed with care.

Actual design of the disk can be varied widely to suit the size and style of the tailstock center.

CLINT MCLAUGHLIN, *Jamaica, NY*

3.28 Double-coated tape holds work

The job called for machining several 3/8-in.-thick aluminum disks of 12-in. diameter to run concentric and true. Surfaces had to be smooth and free of any chatter marks. The conventional method of doing the job on an arbor just didn't work, so the blanks were held to a faceplate with double-coated tape for finish-boring the center hole, turn-

ing the OD, and facing one surface. The parts were then reversed and held the same way for facing the opposite side. Parts came out parallel, concentric, and clean on both sides.

MARTIN BERMAN, *Science Machine Shop, Brooklyn College, Brooklyn, NY*

3.29 Double grooving tool

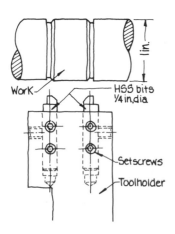

Our shop turns many shafts with retaining-ring grooves that must be spaced accurately to position bearings. To do this job, we have made a special toolholder that accepts two small-diameter high-speed-steel bits ground (on a tool-and-cutter grinder) to the required form. Tool breakage has been reduced from our previous methods, but even if one does break we can insert a new bit in less than a minute and we know that it is in the correct position with respect to the unbroken one.

C P R VITTAL, *Andhra Pradesh, India*

3.30 'Dovetail' jaws grip fixtures securely

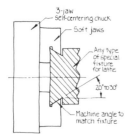

Positive location and safe gripping of special turning fixtures in a three-jaw self-centering chuck can be ensured by boring soft jaws to an inverse taper, as sketched.

For accuracy, the soft jaws must be bored on the machine to be used in production. Then stamp the soft jaws with master-jaw and fixture numbers for future use.

The idea works well with pot-type or box-type fixtures to combine operations on multiple-end parts. Also, it prevents the fixture from pulling out of the chuck even in heavy back-boring or back-facing operations on castings or forgings.

RONALD J ANDERS, *Raleigh, NC*

3.31 Eccentric spade drill for ID-thread relief

The first sketch illustrates a problem we had—the need to open up a deep ID behind a smaller entrance hole. Furthermore, the work material was AISI 4140 with a hardness of Bhn 320, and the machine to do the job was an old, badly worn No. 5 Warner & Swasey.

The original process, quite conventional, was to drill the part to depth, then open the ID with a long boring bar. To avoid chatter with this extremely slender bar, many light passes were necessary, and machining time was excessive. Because of the poor condition of the machine's ways, turret, and other components, the tool would not repeat the dimensions to the tolerances required. And no coolant was reaching the tool.

The problems were solved by a new process based on the well-known fact that a spade-drill will cut oversize if its chisel-edge is not central; I simply designed an offset spade drill together with a holder providing for internal coolant delivery (also using compressed air to help remove chips).

Now a starting hole is drilled only 7 in. deep (rather than 16.7 in.), and this is expanded with a short and rigid boring bar to permit insertion of the offset spade drill. Finally, the special drill is inserted, coolant and air are turned on, and the new tool is fed to total depth (as indicated by the dashed lines in the workpiece sketch).

The drill cuts a diameter equal to twice the distance between the chisel edge and the margin of the wide side, yet the total width of the blade is somewhat smaller than the diameter of the drilled hole through which it must pass.

ENRIQUE M NISENTAL, *Brea, Calif*

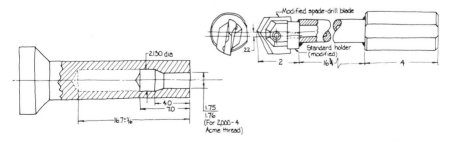

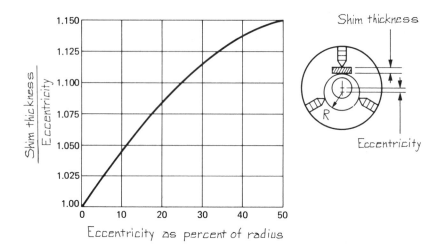

3.32 Eccentric turning in a three-jaw chuck

How thick a shim should you use to pack up one jaw of a three-jaw chuck to produce a given amount of eccentricity? Many people will be surprised to learn that the thickness of the packing is not equal to the desired eccentricity; rather it is somewhat greater. This is due to the fact that the jaws must be opened just a bit wider and the workpiece is displaced slightly with respect to the center of the chuck. In other words, the ratio of shim thickness to eccentricity is always greater than 1. And the ratio varies according to the extent of the eccentricity in relation to the work radius.

The accompanying graph plots the relationship between the shim-thickness/eccentricity ratio and the eccentricity expressed as a percent of radius. The plotted values are theoretically correct, but in practice they are only approximate because they have been calculated on the assumption that the jaws have knife edges and line contact, which, of course, is never true. Nevertheless, the plot provides useful starting points from which to obtain more precise shim thicknesses, if required, by trial and error.

An example will illustrate the use of the graph. What shim thickness is required to chuck a 1.800-in.-dia (0.900-in. radius) workpiece with an eccentricity of 0.200 in.? Eccentricity as a percent of radius (100 x 0.200/0.900) is 22.2%. At this percentage, the graph shows a shim-thickness/eccentricity ratio of about 1.09, which, when multiplied by the 0.200-in. eccentricity, gives a shim thickness of 0.218—or very nearly 7/32 in.

FEDERICO STRASSER, *Santiago, Chile*

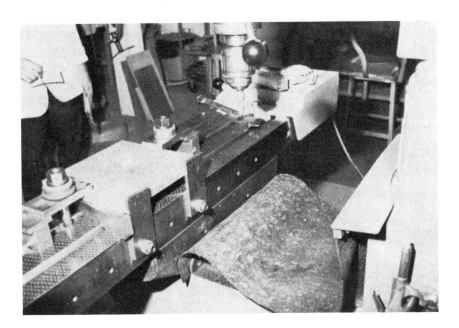

3.33 Edge stops save time in milling setups

Especially useful for setups involving relatively large workpieces made of flat plate is a system of "edge stops" installed along the back edge of table of a vertical milling machine. I simply started in the center of the table and drilled and tapped 3/8-16 holes every 3 in. for capscrews that hold a pair of stops made of flat rock (see photo). They could also be used along the front edge of the table. Using the stops eliminates the need to indicate workpieces that are simply clamped to the table (with or without risers underneath). And they're self-storing.

LAD POCKAJ, *Mayfield Village, Ohio*

3.34 Expanding arbor grips ID

The expanding arbor, or mandrel, shown in the drawing is one of the simplest ways to grip a small workpiece in its ID without danger of distorting or damaging the part. Just cut a suitable length of bar stock, turn the OD 0.0005 in. smaller than the workpiece ID, drill about half-way for about 0.030 in. clearance on the capscrew shank, and then

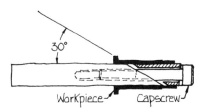

tap drill and tap the rest of the way. Finally, cut the mandrel at approximately 30°, and smooth and deburr the surfaces. With the mandrel held in a chuck or collet, work can be easily clamped and unclamped by tightening and loosening the capscrew.

The same basic principle is also used in our plant to make short expanding plugs when a workpiece has an off-center crosshole to be drilled, reamed, or bored and there is danger of the drill wandering as it attempts to re-enter inside the workpiece ID. Such a plug can be used more than once simply by shifting its position in subsequent workpieces.

ROBERT HORNAK, *Fairfield, Conn*

3.35 Face cam contoured by turning off-center

Here's a relatively simple method we have used for cutting a specially contoured cam on a lathe. It proved to be a low-cost way to produce this type of contour, and it was well within the specifications on the part print.

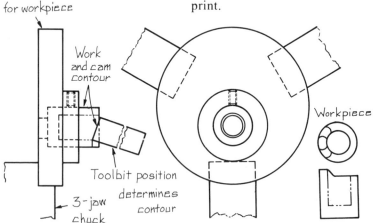

Turning 91

In our case, the cam was made of round plastic rod, which was first set up on-center for boring out the center, thus producing a short, tubular length. The short tube, or cup, was held in a specially made—but simple—eccentric holder that was itself held in a three-jaw chuck, as shown. By angling the broad-nosed toolbit, we set the cam rise precisely as required. The final operation was to cut the part off to desired length, again with it chucked on-center.

MARVIN SMITH, *Sargent Welch Scientific Co, Skokie, Ill*

3.36 Fixture fits lathe, moves to dividing head

The illustrations show a double-duty fixture holder made from a punch holder for a small, 2-ton punch press. The fixture plate is simply 1/4-in. aluminum with a pattern of tapped No. 10-32 holes for clamping the work. This is fixed to the punch holder with four 10-32 capscrews.

The photo shows the workpiece set up for machining on the lathe, after which it was mounted on a dividing head for milling some slots and drilling some holes. Precision of the job is improved by the fact that all operations are done in a single setup on the fixture.

MARTIN BERMAN, *Brooklyn, NY*

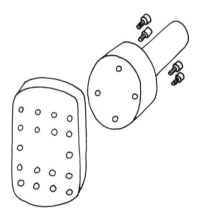

3.37 Flywheel speeds tailstock

When drilling on a lathe, there's always a lot of back-and-forth action on the tailstock quill to clear chips, add oil, etc. To speed up the retraction of the quill, we replaced the standard handwheel with a solid metal disk, on which we mounted the crank handle from the original handwheel.

Now to retract the drill, we just give the

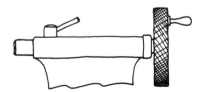

disk a good spin and then stop it when the drill has been withdrawn by the desired amount.

CLINT MCLAUGHLIN, *Jamaica, NY*

3.38 Flywheel spins down chuck

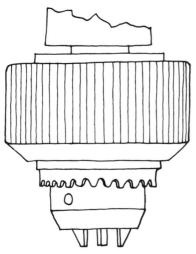

Toolmaking seems to involve a myriad of operations—cutting, filing, drilling, reaming, counterboring, tapping, etc., inevitably alternating with almost as many deburring steps. Anything that saves a few seconds is worthwhile, and the bench drillpress is probably the most widely used machine in the shop.

The drawing illustrates a simple idea that will speed the opening and closing of the drill chuck—and one that is much safer than the not uncommon practice of jogging the start-button with one hand on the chuck. We turned a steel sleeve with a 5/16-in. wall thickness to a slip fit on the chuck. The OD is straight-knurled, and the lower edge must clear the chuck key, but some extension above the chuck (if there's space) adds some useful weight. This sleeve is secured to the chuck with setscrews.

To open or close the chuck, the sleeve is given a twirl, and the added flywheel effect of the sleeve keeps it spinning for a number of turns, thus speeding the opening or closing. You've still got to use the key, of course.

CLINT MCLAUGHLIN, *Jamaica, NY*

3.39 Follow-rest for small jobs

Turning small diameters to any significant length is a vexing task at best. The method described and illustrated here, however, makes it a cinch.

First, machine a block—the one sketched

is for lathes using 3/8-in.-square toolbits—that will clamp in the toolpost as shown in the photo. Next, make a plug of a good bearing material, such as bearing bronze or brass, that's a slip-fit in the follow block and secure it there with the setscrews. Now center-drill, drill, and ream the plug to the desired stock diameter with tools held in collets in the lathe's headstock. Eyeball centering is adequate.

Set the infeed dial on the cross-slide to zero, so you can easily return the block to center if you have to use any other stations on the toolpost.

With the follower on zero, and the stock clamped in a collet and protruding a few inches through the follower, place a 3/8-in. toolbit in the toolpost and tap its cutting edge up against the stock. Now clamp the tool, retract the stock out of the follower, and use a depth micrometer to measure the distance from the outside of the follower block to the tool tip. Back off the depth mike by the amount that the toolbit will have to be moved forward to turn the desired diameter, and reset the toolbit to this dimension.

Now, reinsert the stock in the follower, clamp the collet, and turn the stock to the required diameter.

I have used this setup to turn 1/8-in. stainless down to 0.075 in. ±0.005 in. for a length of 2 5/16 in.

STEVEN R CARRIER, *Bellingham, Mass*

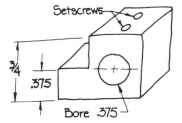

3.40 Four-jaw chucking made even easier

Truing up pieces in a four-jaw chuck can be time-consuming and frustrating. But I've got a way that speeds the task. I merely mounted four inexpensive steel scales on

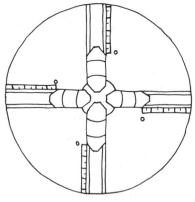

the face of the chuck next to the jaws. The job is done as follows:

First true up a piece of round stock (preferably the smallest diameter the chuck will accommodate) with an indicator. Then clamp the scales to the face of the chuck with the zero point at some distinguishing mark or witness mark on the jaws. I used the edge formed by the 45° angle at the end of each jaw. If the scale zero is placed at the outside of the jaw with the chuck at its minimum diameter, the witness mark will function over the full range of the jaw.

The arrangement not only helps in quickly centering work but also helps on eccentric jobs because the offset can be quickly determined and almost preset.

EDWARD J IDE JR, *Waterloo, NY*

3.41 Grinding chuck jaws

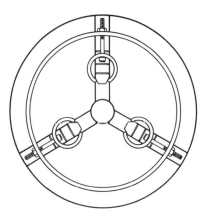

When running a lathe job that requires accurate centering of the work in a chuck, it's a good idea to grind the jaws. This should be done with the jaws set at the workpiece diameter, and the chuck should be set up so that the jaws can be tightened against the scroll.

To do this, we use one large ring and three small ones, the latter three cut to clear the portion of each jaw to be ground, as shown. Three long screws provide a wide range of adjustment latitude for jaw position.

ERNEST JONES, *Yorktown Heights, NY*

3.42 Gripping threaded work

Sometimes you have to turn down the head of a bolt or do some other lathe operation on a threaded part and you just can't avoid gripping it on its threaded portion. Such work can be held securely and with no danger to the threads by simply turning down a suitable nut (or a pair of nuts) to fit the ID of a collet, then slit it with a hacksaw so it will pinch the threads and hold the bolt securely and safely for turning.

TONY SIRACUSA, *Sewell, NJ*

3.43 Holding square stock in a three-jaw chuck

When tolerances permit and switching to a four-jaw chuck is not practical, a quickly made split bushing can be used for chucking square stock in a three-jaw chuck. Just make the bushing ID slightly larger than the diagonal measurement of the square stock, as illustrated. The method can also be used for center-drilling the tailstock end of longer lengths.

RAYMOND SIMET, *Wayne, NJ*

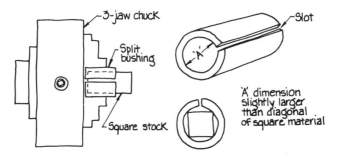

3.44 ID grooving with a Woodruff cutter

When I have to cut standard-width grooves in an ID (for snap-rings, O-rings, etc) while machining a part on a lathe, I use a standard Woodruff key-seat cutter. The Woodruff cutter, of course, must be smaller in diameter than the bore, and it's mounted in the toolpost as a boring bar would be. One tooth must be set on center height, and this one does all the cutting.

FRANCIS J LEE, *Warminster, Pa*

3.45 Internal driver eliminates dog for turning

This driver is ideal for turning tubular work or parts with drilled or bored holes in the headstock end that can be supported by a tailstock center at the other end. No lathe dog is needed, a good cut can still be taken, and the entire length of the part can be machined.

Machine a chucking shank on one end of a piece of steel bar and a concentric 15° taper (30° total included angle) on the other end, as shown. Then mill three longitudinal slots spaced at 120° on this taper. Slot width should fit 1/4-in. or 1/2-in. standard toolbits, depending on the size driver

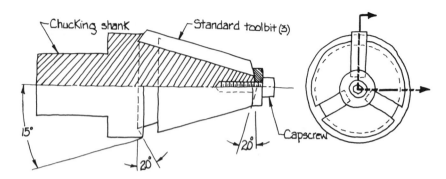

being made. Now turn a groove near the big end with an undercut that matches the grind on the ends of the toolbits (preferably 20°), and machine a thick washer with a similar matching angle for the small end. These features will retain the toolbits in the driver. Next, insert the three toolbits, and tighten the washer with the capscrew you've installed in the end of the driver.

To use the driver, chuck it in the lathe and mount the work between it and a live center in the tailstock. Apply pressure from the tailstock, and the OD is ready for turning. And when the toolbits become worn or chipped, they can easily be removed and replaced.

LESTER RUTHSTROM, Houston, Tex

3.46 Keep a cool hand while polishing shafts

It's often necessary to polish long shafts in a lathe to get a uniform size or finish. The job is typically done by holding a piece of abrasive cloth on the rotating shaft manually. The problem is that you quickly feel the heat

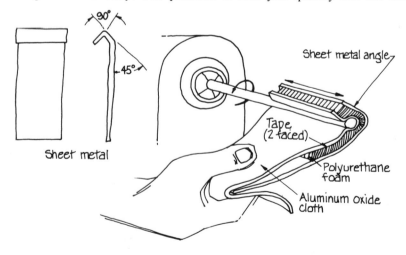

generated—uncomfortably—and maintaining a precisely uniform diameter over the entire length is difficult.

A backup block of wood or metal is often used to apply the abrasive cloth somewhat like a file. This helps to maintain diameter consistently, but the line contact between the metal workpiece and the hard backup block quickly destroys the effectiveness of the abrasive and slows stock removal.

The simple shaft-polishing tool in the sketch solves both problems. It's just a 2-in. width of 1/16-in. sheet metal bent approximately as shown. A 2-in. square of 3/8-in. or 1/2-in. polyurethane foam is held inside the angles with double-coated tape, and a length of 2-in.-wide abrasive cloth is laid in this with the grit side up. Thumb pressure from one hand clamps the cloth, and the other hand guides the tool back and forth over the work.

The tool has to be used with care. Although it seems to generate less heat than other methods, it seems to remove stock more quickly—so watch that workpiece diameter. As the cloth wears, it's quickly advanced by letting it slip slightly beneath the thumb that secures it.

ART DRUMMOND, *Walworth, NY*

3.47 Lathe attachment auto-feeds for threading

The drawing shows an attachment we clamp onto the compound rest of a lathe to provide automatic feed between passes in single-point threading operations.

Base of the attachment is a rectangular steel block bolted to the compound. The block is drilled and tapped for a shoulder bolt on which the toolholding lever pivots and which also serves as the operating handle. (To simplify the drawing, a few details, such as the slot and setscrews holding the threading bit in place, have been omitted.) Two pins are pressed into the side of the base-block, one below the front end of the lever to take the cutting force, and the other at the rear to act as a limiting stop when the lever is pushed down.

A ratchet wheel and a knurled knob are added to the feedscrew as shown.

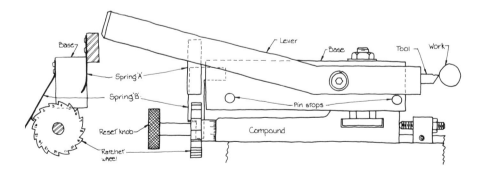

The ratchet is actuated by a flat spring secured to the lever, and a second flat spring on the other side of the base-block prevents reverse rotation.

To cut a thread, the attachment is set up as shown, and the first cut is made in the usual manner. At the end of that pass, the lever is depressed to lift the toolbit out of the cut and also to increment the ratchet wheel to advance the compound's feed. The carriage is then returned to the starting position, and the lever is lifted to bring the toolbit back into cutting position. This procedure is repeated until the thread has been cut to full depth, at which point the anti-reversal spring is lifted out of the ratchet wheel and the compound is reset.

If desired, an adjustable stop can be added to the compound as shown to aid in setup. Reset is facilitated by the numbers stamped on the ratchet wheel, and the number of teeth should be determined by the size of the lathe and the pitch of the feedscrew. A spring washer under the head of the shoulder-bolt pivot adds friction so the lever is held in its desired position. When conditions permit, two strokes of the lever can be used to double the depth of cut. And the ratchet wheel and knurled knob need not be removed from the compound when the attachment itself is removed.

CLINT MCLAUGHLIN, *Jamaica, NY*

3.48 Lathe cuts eccentric reliefs on form cutter

The drawing shows a cutter-backoff attachment we built for use on a lathe. Although it appears rather complicated, it is fairly simple to make because it will do accurate work even if it is itself not built to high accuracy.

Part "A" is an L-shaped steel plate with its base leg drilled for attachment to the side of the lathe saddle (in the tapped holes put there by the builder for mounting a follower rest). The upper leg of the L has a bronze bushing for a shoulder-screw that goes through spacer "B" and into the T-nut "C." The T-nut can be adjusted up or down in a mating T-slot cut in bar "D." When the shoulder-screw is tightened, it locks the T-nut in place, but permits bar "D" to pivot.

On the opposite face of bar "D" is a lengthwise T-slot that accepts a nut to allow bolting the cutter blank in place. A setscrew locks this T-nut in position, so that it does not shift each time the cutter is loosened to index it to the next tooth. The back end of bar "D" rests on crankpin "E" on crank assembly "F." Shank "G" of this assembly is held in the lathe chuck. Pin "H" at the front end of the bar is used for indexing the cutter being made.

To make a backed-off cutter with this setup, the blank is first turned to the desired diameter and the slots or gullets are cut. Accurate spacing of these slots is not vital because each cutter tooth is backed-off with the index pin in

Turning

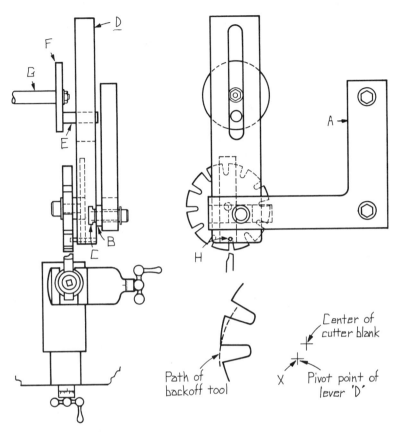

the adjacent slot. The blank is then mounted as shown in the drawing. As the lathe rotates, crankpin "E" cams the back end of "D" up and down, causing the cutter blank to oscillate up and down about eccentric point "X" so that the lathe tool swings across the cutter tooth as shown by the dotted line.

By adjusting the positions of the two T-nuts, the relationship of the two pivot points changes, modifying the cutter path. The lathe tool, which of course can have a special form to impart to the milling cutter being produced, is fed in until the tooth is properly backed off, and the process is repeated for each tooth to the same final position.

CLINT MCLAUGHLIN, *Jamaica, NY*

3.49 Lathe dog grips on narrow flanges

The sketch shows a lathe dog I use when workpieces with narrow flanges have to be

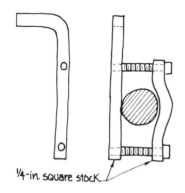

turned between centers. Conventional lathe dogs are just too clumsy for such work, and the setscrew they use for clamping often mars the work. The dog shown was simply bent up from 1/4-in.-square cold-rolled steel, and was drilled and tapped for the clamp screws.

MARTIN BERMAN, *Science Machine Shop, Brooklyn College, Brooklyn, NY*

3.50 Lathe-tool height setter

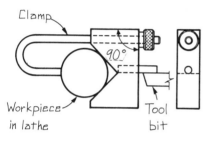

We can live with a toolbit being set slightly below center, but a toolbit set above center can cause problems. Also, if it's a form-tool and it's stepped, the part will not be dimensionally correct if the tool is not dead on. To help set lathe tools just right, we made the simple V-block device illustrated.

The V-block handles diameters up to 2 in., and a 1/4-in. bent rod threaded for a knurled nut is used for clamping. The pin indicating center is pressed in place.

Three methods can be used in toolsetting: (1) All surfaces are ground so that a square can be used to align one side vertical. (2) A bubble level can be placed on top to assure that surface is horizontal. Or (3) a dial indicator can be used to set the pin precisely on center.

In all cases, of course, the toolbit must be set flush with the underside of the pin when the toolholder is locked up.

ERNEST J GOULET, *Middletown, Conn*

3.51 'Machining' foam plastic

Have you ever wanted to machine a piece of Styrofoam relatively accurately? It's quite easy if you simply use a hot tool.

I had to make a core for holding NC tapes so they'd go through a high-speed tape reader properly. The job was done by chucking a piece of Styrofoam in a lathe and using a soldering iron for a cutting tool. A small Ungar iron with a No. 8 or hotter tip works well. It just peels the Styrofoam off as the carriage moves along the bed.

LEO U SILVER, Santa Barbara, Calif

3.52 Magnet marks drill depth

When drilling from the carriage of a lathe, place a small magnet on the ways so that it will be moved along as the tool is advanced. This will mark the maximum advance of the drill when it's necessary to back out for chip clearance and will help prevent ramming the drill to the bottom of the hole upon re-entry. You can do the same thing on drill presses and other types of machines, as well.

MARTIN J MACKEY, Parma, Ohio

3.53 Magnetic spacers ease chucking jobs

I recently had the job of machining some 12-in.-dia rings that were 3 in. long. The rings had to be bored and the ends faced off parallel. To bore the rings in a three-jaw chuck, they had to be kept clear of the jaw faces (so the boring tool wouldn't hit the jaws). Spacers behind the ring had to be wired in place, or taped, until the ring was locked in the chuck, and then they had to be removed before machining so the tool wouldn't hit them. It was a tedious waste of time.

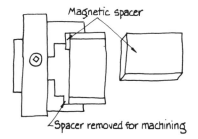

I speeded the job with several magnets, nominally 1/4 in. x 1 in. x 2 in., which I ground all to the same thickness. The magnetic spacers easily stick to the chuck jaws and are easy to remove after the chuck is tightened on the ring. And they saved a lot of time on that particular job.

JOHN URAM, Cohoes, NY

3.54 Morse-taper toolholders for a lathe turret

To substantially increase the production of engine lathes equipped with full cross-slides,

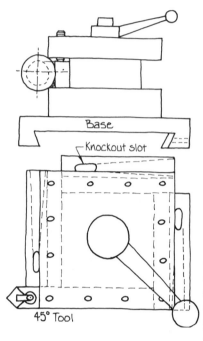

the customary rear toolpost can be replaced with a square turret. And with this added toolholding capacity, four of the most commonly used tools can be mounted there ready for instant use.

For toolroom or general use, I use a 45°-lead, square-insert toolholder for facing and chamfering in one position. At the other three positions, I clamp special holders consisting of a Morse-taper sleeve welded to a rectangular clamping shank, as shown. One of these mounts a center-drill, the second a Jacobs chuck, and the third is open for whatever size drill may be required.

These Morse-taper holders must obviously be carefully made so they are vertically on center and truly parallel to the spindle axis. Horizontal centering for drilling operations is achieved by mounting the turret so that when the cross-slide is run out to its maximum limit, or to a special stop, the tapers are on center. For production jobs, the turret can be set closer for faster positioning, and a stop can be set for position.

EDWARD J IDE JR, *Ide Machine Co, Waterloo, NY*

3.55 Part-holder for cutoff

The tooling shown in the sketch is used on our turret lathes to cut off parts without leaving the usual central burr.

The unit consists of a length of barstock turned down to fit into a needle bearing gripped lightly in the turret, and with a ball thrust-bearing between a shoulder on the shaft and the turret face. The outer end of this shaft is bored slightly oversize to receive the workpiece and is internally grooved for a pair of rubber O-rings. The shank end is drilled for a part-ejecting pushrod.

Before the workpiece is cut off, this tool

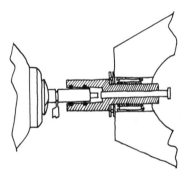

is pushed over the part (on the O-rings) and starts it spinning. When the workpiece is then cut off in the usual manner, the tool both supports the work and keeps it spinning until the cutoff tool is through. In some cases it may be advisable to make the tool heavier, so it keeps the work spinning a little longer.

CLINT McLAUGHLIN, *Jamaica, NY*

3.56 Parts catcher for chuckers

Since our manual and automatic chuckers are not equipped with part chutes and picking up the parts after cutoff can be a nuisance, we devised the simple parts catcher shown in the sketch. We simply drilled and tapped a hole at a 3 degree angle in a toolblock to accept a 10-32 socket-head screw. Plastic screws are ideal for parts with fine internal threads or critical internal finishes. Size of the screw, of course, can be varied to suit the parts being produced. We use the method to catch spacers, retainers, cells, and other ring-shaped parts.

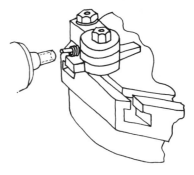

As the illustration shows, the catcher moves into the center hole in the part, and as the cutoff slide advances the part simply drops onto the catcher. It isn't always necessary to remove the parts one by one as they are machined. This depends on the length of the part and the length of the screw. We recently machined 1000 aluminum spaces, 3/4 in. OD, 5/8 in. ID, and 0.040 in. thick, and we let 25–30 parts accumulate on the catcher before removing them.

WILLY COLEMAN, *Whitestone, NY*

3.57 Plug speeds chuck, steady-rest setup

We do a lot of work on long shafts, and we've found a good use for the stub

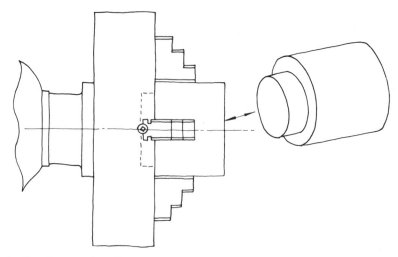

ends. On the popular sizes, we turn up a piece as shown in the drawing, with the smaller diameter a slip-fit in the through-hole of our four-jaw chuck. The plug is then used to preset the independent jaws of the chuck. The plug is inserted in the chuck hole, and the jaws are tightened equally against it. The four jaws are then all backed off a half-turn, the plug is removed, the workpiece is inserted, and the jaws are tightened a half-turn.

We make the plugs long enough to project out beyond the jaws. This way we can also use the plug for setting up the steady-rest jaws, after which we move the steady out to the desired position on the lathe bed.

CLINT MCLAUGHLIN, *Jamaica, NY*

3.58 Precise centering in a floating toolholder

The inability to achieve the exact centering of a tool held in a floating toolpost can cause hours of wasted trouble-shooting in building lathe setups. We use a very quick and simple technique for providing precise center location in such cases.

The tool to be mounted is held in the spindle of the machine in a collet of appropriate size. The toolholder, mounted in the turret, is traversed over the tool in the spindle. Then it's simply a matter of tightening the floating holder onto the tool while it is being held firmly on center in the collet. Then just go on to finish the rest of the setup.

JOHN KARPOWICZ, *Philadelphia, Pa*

3.59 'Puppy' dog for small jobs

Most tool cribs seem to have plenty of lathe dogs suitable for all sizes of work from about 1/2 in. diameter on up, but when you have to turn a 1/4-in. shaft on centers even the smallest lathe dog is too big. The accompanying sketch shows a simply made dog for such miniature work.

It consists of 1 1/2-in. length of 3/16- x 1/2-in. aluminum with a hole to suit the workpiece drilled and reamed about 1/2 in. from one end. The end is slit with a saw into this reamed hole, and provision is made for a 4-40 socket-head cap screw for tightening this clamp. The other end is drilled and tapped 8-32 for another cap screw (with lock nut) with which to drive the dog.

Another advantage to these lathe dogs, which, of course, can be quickly and easily made to any dimensions required by the job, is that they do not mar previously finished surfaces on the work.

JOHN URAM, *Cohoes, N Y*

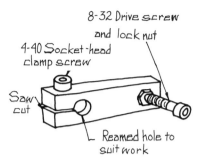

3.60 Quick way to accurate depths on a lathe

In the course of drilling and tapping on an engine lathe, I occasionally run into jobs in which depth is critical. Even though some tailstocks are graduated in thousandths, a more accurate method can be used. Just tighten a C-clamp lightly on the tailstock quill of the lathe and mount a long-throw dial indicator on the tailstock casting with a magnetic base. Then adjust the setup so the indicator reads zero at maximum tailstock advance. It's simple to set up—and precise to use.

JIM FITZGIBBON, *Oklahoma City, Okla*

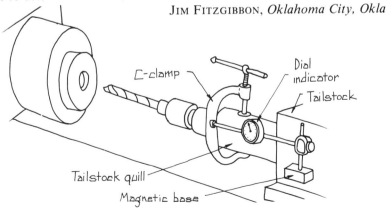

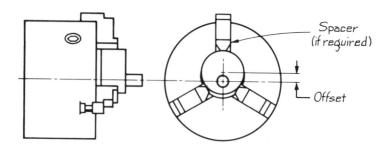

3.61 A quick way to chuck eccentric turning jobs

By skipping one or more teeth during insertion of the jaws in a three-jaw scroll-type chuck, round stock can be offset almost any desired amount for turning or boring eccentric jobs. A small spacer can be added under one of the jaws if the pitch of the scroll doesn't give the desired offset to the workpiece in the machine.

PAUL CAPUTI, *Bloomfield, NJ*

3.62 Rebuild recycles old long-bed lathe

An old, large, long-bed lathe was giving us problems because of its slow revs, headstock condition, and wear on the leadscrew and bed. Among alternatives being considered were scrapping the old thing or overhauling it at considerable expense.

However, it was noted that years of chucking operations had restricted most of the wear—on both leadscrew and ways—to the first 2 ft of the machine; the original scraping marks were still clearly visible on the remaining portion of the bed. Then we hit on the idea of building a simplified headstock, mounting this in front of the old one, and coupling them together so that all existing drive mechanisms could be used.

The new "box" is made of 15-mm (0.6-in.) welded steel plate and the base of 20-mm (0.8-in.). This contains the new spindle mounted in heavy-duty, precision tapered-roller and ball bearings, a coaxial input shaft externally coupled to the old chuck by means of a welded flange and three bolted spacers, and a countershaft gearing arrangement that doubles spindle rpm. Automotive type helical gears are used—low-cost and quiet—running in an oil bath.

The doubled spindle speeds and heavy-duty bearings now permit the use of modern insert-type cutting tools. We have to remember, of course, to set feeds to half of those desired because of the double spindle speeds. Effectively, we swapped 2 ft of bed length for a machine that is essentially in "as new" condition. And an additional benefit is that we have gained a new method of

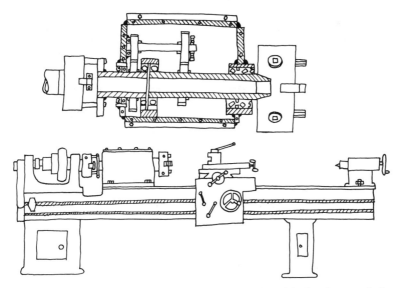

chucking long bars—using both the new and the old chucks much in the manner of oil-country lathes rather than with a steady rest. Since the original job was done about two years ago, we've modified two more machines for other shops.

B IVALDI, *Turin, Italy*

3.63 Recycled center-drill makes radius tool

Worn-out or broken center-drills can be converted into good radiusing tools for turning fillets on a lathe. The toolholder shown in the sketch, which is made of cold-rolled steel, works well for me. As well as cutting accurate radii, the tool is also useful for producing fine surfaces on finish cuts.

ED ZATLER, *Westland, Mich*

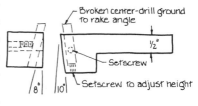

3.64 Removable steadies support big work

These simple supports are made of 1/2-in. x 3/4-in. rectangular bar welded up as shown. A clearance hole in the base leg allows clamping to the face of the lathe chuck, and a 1/2-13 tapped hole at the outboard end allows it to be tightened against the work.

The basic purpose was to hold tubular work that was too big for the

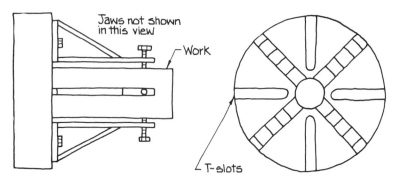

available steady-rest for machining the ID. The steadies hold the work firmly, and the outboard bolts allow it to be centered.

GARY A STEPHENS, *Taylorsville, Ky*

3.65 Removing work from threaded mandrels

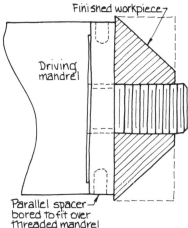

The best method for turning some internally threaded workpieces is to machine the ID first, and then complete the OD with the part threaded onto a mandrel held in the lathe's spindle. Typically, however, many such parts have external configurations that make them difficult to remove—either because they can't be gripped conveniently or because the surface finish must not be damaged.

A simple method for solving this problem is the use of a parallel spacer bored out to fit over the mandrel's threads and wide enough so that several holes can be drilled into its OD to accommodate an adjustable-pin spanner wrench.

Turning the part, of course, drives it tightly up against the spacer—as it would against the mandrel without the spacer. But rotating the spacer counter-clockwise with the spanner will also rotate the part. And as soon as slight movement takes place the workpiece can be freely unscrewed from the mandrel.

In machining the mandrel, it's desirable

to relieve as much of the back surface as possible, and to apply a film of grease on that surface. This will help reduce the friction for removal.

JACK MOORE, *Yorktown, Va*

3.66 Rest pads aid chucking thin disks in lathe

Holding parallelism of large-diameter rings or disks that are thinner than the depth of the chuck-jaw steps can be accomplished quickly and easily with these rest-pad locators bonded to the face of the chuck.

The three locators are machined from aluminum, as shown, with a large base diameter for ample bonding area and a small pad diameter both to minimize cutting forces while truing them and to provide essentially point contact with the work. Overall length of the pads is left about 0.005 or 0.010 in. oversize during initial machining so that the pads can be trued accurately after they have been bonded to the chuck (either with or without the jaws in place, depending on the specific setup).

Of course, when using any of the cyanoacrylate "super" glues, the bonding contact areas must be thoroughly clean and free of oil. A fine-grit abrasive paper or steel wool followed by wiping with alcohol or other solvent will do the job.

The pads can easily be knocked off, or they may be left in place if they do not interfere with other operations.

CHARLES L VECCHIATTO, *El Toro, Calif*

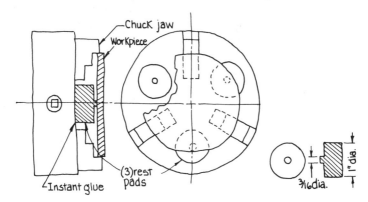

3.67 Rotary table helps turn a concave radius

Our shop recently had a job that called for a 12-in. concave radius on the rim of a wheel assembly. The job would ordinarily be done by numerical control

or on a tracer lathe, but we have a small job shop and all of our equipment is manually controlled.

Width of the steel rim was 3 1/2-in. and, needless to say, a ground high-speed-steel form-tool howled and chattered—producing a completely unacceptable surface finish.

We finally solved the problem with the setup shown in the photo, which involved a rotary table bolted to a ground plate mounted on the lathe cross-slide. A cutting tool was mounted on the rotary table and set to the required 12-in. radius. Despite the rather large tool overhang necessitated by the required radius, several passes over the workpiece produced an excellent surface finish well within the required tolerance of 0.002 in. on the concave radius and the wheel diameter.

ROGER K CARLSON, *Schererville, Ind*

3.68 Self-locking lathe arbor

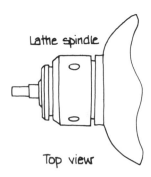

Use of a self-locking arbor can greatly speed and simplify workpiece exchange when it's necessary to turn a quantity of parts that must be gripped internally, but which block access to the expansion screw of an expansion arbor. Such parts might be blind bushings, capped sleeves, certain collars, etc.

To make a self-locking arbor, turn a piece of stock to a sliding fit in the workpiece and mill a horizontal setup in one side as shown on the drawing. Depth of the step must be such that when the workpiece is rotated toward the operator, a 1/8-in. or 3/16-in. pin will roll into the deep area and free the work. Rotating the part in the other direction will then roll the pin into the shallow area and lock the piece by a cam-like action to resist cutting forces.

ALBERT T PIPPI, *Baltimore, Md*

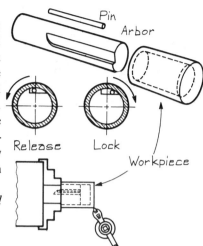

3.69 Setup for soft-jaw boring

The illustrated adjustable tool, made out of 1 1/2-in. hex stock, is used for preloading a lathe chuck for boring soft jaws to size. Its use eliminates the need for keeping a collection of various-sized disks for the purpose—which rarely seems to include one that's close to the diameter of that new workpiece.

ED ZATLER, *Westland, Mich*

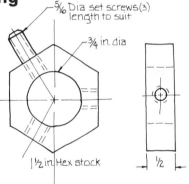

3.70 Setup positions tool accurately on lathe

It was necessary to have an inexperienced operator run an old lathe—the kind with a small, coarsely graduated dial on the crossfeed. And to make matters worse, a large number of pieces had to be turned to a close tolerance.

The problem of final positioning of the toolbit for the finish cut was simplified by mounting a dial indicator in the tailstock, as shown in the sketch. The indicator was

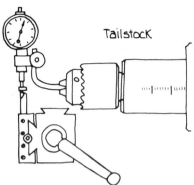

fitted with a custom-made carbide tip with a flat surface.

The ram or the entire tailstock can then be positioned to leave sufficient room for loading and unloading workpieces, yet minimize the carriage travel necessary for setting the toolslide each time.

JOHN URBAS, *Cannon Tool Co, Canonsburg, Pa*

3.71 Simple fixture simplifies eccentric turning

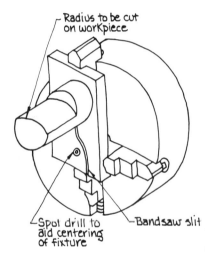

Here's a simple setup for machining a radius on a part that's larger than the part diameter, a situation that defeats normal use of the four-jaw chuck. The split clamp holds the work, while the chuck jaws hold the split clamp and also supply the clamping force to secure the work.

The clamp is just a suitable piece of plate bored to fit the workpiece OD, spot-drilled for locating the center of the radius to be machined, and bandsawed to produce the split. Just make sure in the bandsawing operation to avoid putting the slit in the area that will be clamped by the bottom jaw.

The setup was meant for use with a rotary table on a milling machine—the workpiece radius is milled with multiple circumferential passes. But a finishing cut can be made with the same setup at slow speed on a lathe.

RAYMOND B HARLAN, *Wayland, Mass*

3.72 Simple gage checks bore size in lathe

Boring out a few workpieces is a common engine lathe operation. But it's typically a precision operation, and operators—especially new ones—are very

Turning 113

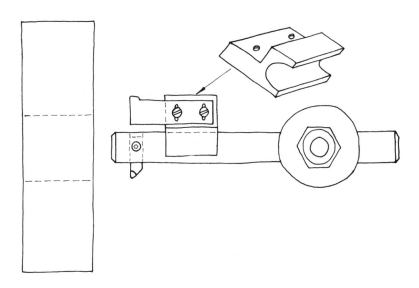

cautious in this type of work to prevent boring oversize and ruining the job. To expedite such jobs, we made up a simple adjustable gage that slides along the boring bar for a quick check on the bore's diameter.

This consists of a brass or aluminum block machined to slide on the boring bar with an adjustable blade screwed to a step in the block as shown. To use the gage, it's first set up as shown and the blade is adjusted so that the dimension from the tip of the boring tool to the tip of the blade is 1/64 in. less than the desired finished bore diameter.

Now, as each cut is started, the gage is manually slid along the bar until the blade contacts the end surface of the work. In this position, it gives a quick indication of the amount of stock still to be removed. And once the gage enters, the finish cuts are taken and a bore-gage is used for final, precise measurement.

CLINT MCLAUGHLIN, Jamaica, NY

3.73 Simple gage simplifies steady-rest setup

This simply made setup tool greatly facilitates the setting of steady-rest pads or rollers to the proper diameter. A broken drill or reamer with a suitable taper shank and a short slug of barstock provide the basic raw materials, which should be machined as shown.

To use the tool, first measure the diameter that is to run in the steady, set an inside micrometer to half this diameter less the 0.250-in. radius of the setup tool's spindle, mount the ID mike in the collar with the thumbscrew (making sure it's in contact with the 0.500-in. spindle), slide the tailstock up to the

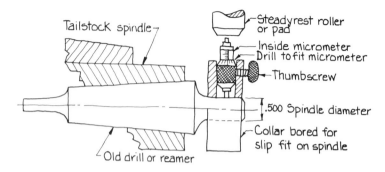

steady-rest, and use the micrometer setting to adjust the pads or rollers one at a time.

STEPHEN VALAN, *Gillette, NJ*

3.74 Simplify four-jaw chucking

The independent adjustment of a four-jaw chuck makes it possible to mount work in a lathe with high precision, and, of course, it permits chucking of irregularly shaped workpieces. But initial setup of the workpiece within the range of the dial indicator used for final centering—especially with heavy castings and the like—is often cumbersome.

I've found a way to facilitate this job in many cases, depending on the workpiece shape. What I have done is to mount a three-jaw self-centering chuck on a rotating arbor that fits the lathe tailstock. Then I grip the workpiece in this—and most workpieces seem to allow it—and then move it into the four-jaw chuck on the spindle.

The method allows me to carefully locate the part in relation to the jaws, and to place shims under them, while rotating the part and tightening the four-jaw chuck. Final adjustments, of course, are made according to an indicator.

JAMES R WOOD, *Columbus, Ind*

3.75 Single setup turns OD and ID at both ends

Both internal and external turning and threading operations on both ends of thin-walled cylindrical pressure vessels are done in a single setup with the illustrated workholding arrangement developed in our plant.

Basic elements of the device are a pair of three-jaw self-centering chucks with pie-shaped top jaws (to avoid distortion of our thin-walled workpieces). As shown, these grip on the ID at both ends while providing adequate clearance for the internal operations. The tailstock-end chuck is mounted on roller

Turning

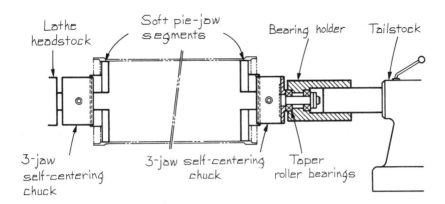

bearings in a holder clamped on the tailstock spindle, so that this chuck rotates with the work.

A variety of top-jaws have been machined to accommodate workpieces of different ID.

A V HANUMANTHA RAO, *Trivandrum, India*

3.76 Sleeve stabilizes boring bar on turret lathe

We run a variety of different-size rollers and bushing on our Warner & Swasey 2AC chuckers, which means that we're constantly boring out jaws to ensure good concentricity. On the smaller jobs we usually use a 1 3/4-in. boring bar inserted in a sleeve and then into a retractable boring head. With the head moved all the way forward on the pentagon and the boring bar set for proper depth, we sometimes find the bar hanging out 5 in. or more. Regardless of the rpm or feed being used, the boring bar is just not sufficiently rigid, and deflection is a problem.

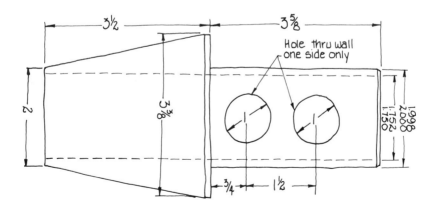

The sleeve in the drawing, however, will add some rigidity to the boring bar and will help to eliminate deflection problems and scrapping the job.

DAVID P ZECHMAN, *Huron, Ohio*

3.77 Slitting saw makes narrow grooving tool

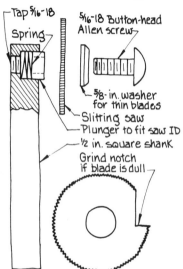

You can use broken slitting saws—or even new ones—for turning narrow grooves in a lathe. Such saws come in diameters from less than 1 in. to several inches and in a range of thicknesses from about 0.005 in. up. I have used this method for years with a toolholder (see sketch) made up to fit an Aloris toolblock holder. If the sawblade is dull or broken, it's a simple matter to grind a new cutting edge, as shown, similar to a circular form-tool as used in the screw-machine industry.

RON STANWICK, *Englishtown, NJ*

3.78 Soft jaws hold collet pads for chucking

The practice of boring out soft jaws every time a new run of parts is to be turned from cut-off slugs in sizes up to 2 1/2-in.-dia rounds or 3-in. hexes can result in a very large inventory of special jaws for different diameters and part lengths.

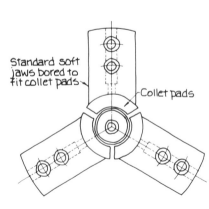

More costly than the inventory problem, however, is the time and effort needed to machine these soft jaws—especially if undercuts are necessary to clear burrs on the slugs or if secondary milling is required at the gripping edges to allow the jaws to close on small diameters. Then there's the fact that soft jaws tend to get "beat up" with repetitive use, leading to excessive runout, bellmouthing, and similar damage that results in reduced work accuracy.

These problems, which probably increase the total costs of using soft jaws more than you think, can be avoided by using standard, off-the-shelf collet pads for different round and hex sizes. A set of soft jaws can be bored to adapt to the collet pads, and then be carburized or hardened. The pads, of course, simply screw in place.

Chucking length can be accommodated with either a fixed-length stop or an adjustable one. Using a stop smaller than chucking diameter eliminates the need for undercutting, and since the OD of all collet pads (of the same type) is the same, irrespective of ID, there's no need to mill the jaws for smaller diameters. The smaller drawing shows a stop plate, which bolts to the face of the chuck, for holding workstops.

It's a fast, easy, and efficient method for holding slugs in a lathe. Changeover for a different size or shape (round/hex) is quick—as is any switch of depth stops. The problems of runout, bellmouth, and worn jaws are largely eliminated. And machining of new jaws for every diameter and length is avoided.

RAJEEV KRISHNA, *Chicago, Ill*

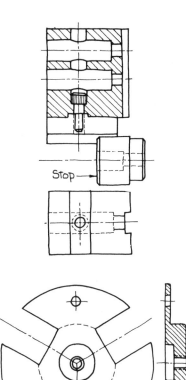

3.79 Sometimes a mill is a lathe

When it's necessary to turn or face the end of a rod, especially one of small diameter, I often find it preferable to do the machining in a vertical mill. The workpiece is held in the spindle chuck, and the toolbit is held in a milling-machine vise on the table.

For facing, the spindle is set and locked in position, the table in-out is adjusted to bring the toolbit on center, and the tool is fed by traversing the table to the right.

For turning a smaller diameter on the end of the rod, the setup is similar, but the

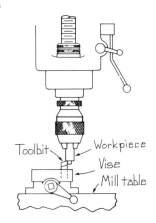

spindle is not locked, and the spindle stopnuts are adjusted to control the length (depth) to be machined. Again the table position is adjusted to put the bit on center, but this time it's also set for the desired depth of cut and then locked. Tool feed is then accomplished with the vertical spindle-feed lever.

One of the reasons for doing such operations on a mill is that higher spindle speeds are generally available, which yields a better finish on the work and lets you do the job in less time.

RICHARD L ROGERS, *Warren, Pa*

3.80 Special holder for radius-turning tools

One recent job I had to do was to machine a radiused groove with a diameter of 0.109 ± 0.001 in. Rather than grind the radius on a toolbit, I made up this simple tool, which works beautifully. The holder is made of 1/2-in.-sq cold-

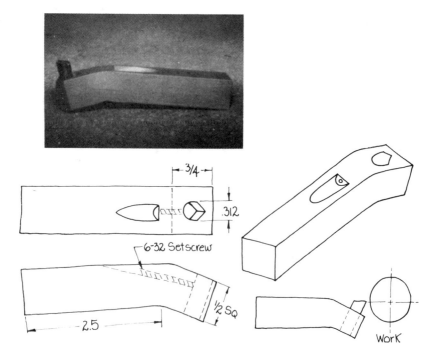

rolled steel, bent to 18°. A 5/16-in. hole was drilled through this, as sketched, and a 90° V was broached at the front of this holder. The body was then relieved with an end-mill, and it was drilled and tapped for a 6-32 setscrew to hold the actual cutter, which was made of a short length of hardened drill rod. The cut was clean, chatterless, within tolerance, and required no polishing.

Now I can produce a range of accurately radiused grooves simply by making up the required "insert" from hardened drill rod, hardened drill blanks, or even broken high-speed-steel tools.

RON STANWICK, Somerville, NJ

3.81 Spider centers large pipe for turning end

The workpiece was a 24-in. length of 14 in.-dia pipe with a 3/4-in. wall thickness. The job required facing the ends, turning the OD, and cutting a number of grooves in the surface. I put the job in a four-jaw chuck (with jaws reversed) on a 20-in. lathe and supported the work with tailstock pressure on a wooden 2 x 4 across the end while I indicated it closely enough and tightened the chuck.

But the job couldn't be machined without better tailstock support. So I made up a spider for use with a live center in the tailstock.

For the spider, I used 1-in.-thick aluminum plate turned to a 6-in. OD. To ensure concentricity, this was center-drilled with a large center drill in the same setup. On the OD, I drilled and tapped six equally spaced holes 1/2-13 by 2 in. deep (the sketch shows eight, which would give better support). I then cut six suitable lengths of 1/2-13 threaded rod and machined flats on each for an adjusting wrench. Locknuts were added, and the rods were screwed into the plate.

The assembly was inserted into the end of the workpiece, leaving clearance for facing the end, and the spider was centered in the pipe with the aid of an indicator on the OD of the plate. Locknuts were then tightened, and the tailstock center was brought in to provide the support for machining the job.

JOHN URAM, Cohoes, NY

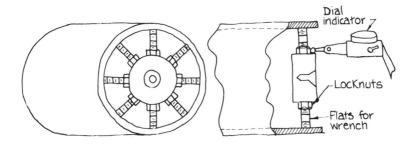

3.82 Split expansion arbor speeds turning job

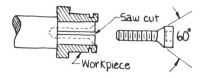

Sometimes you get a lathe job, typically 25 pieces or so, that calls for facing both ends, boring an ID, turning two or three diameters, and plunging a groove or two. To finish each one complete would be excessively time-consuming because of the several necessary tool changes. I have found that a simple split arbor made of aluminum will almost cut the time in half.

For the part sketched, which has a 0.750-in. bore, I faced one end and then bored out each piece in the lot. Then I mounted a piece of 1-in. aluminum barstock in the bench lathe, drilled and tapped it 1/4-20, center-drilled it with a large center-drill, split it with a saw cut, and finally turned its gripping diameter to 0.749 in. Into the tapped hole went a 1/4-20 socket-head capscrew on which the head had been turned to a 60° angle (as shown).

With the arbor still chucked, each piece of the entire part run was then faced on the other end and the ODs were turned, then the tool was changed and all parts were grooved to complete the job.

The arbors are worth saving, because they can subsequently be turned down to grip in smaller bores for other jobs.

JOHN URAM, *Cohoes, NY*

3.83 Sprag arbor is quick, convenient

Here's a quick and easy method for gripping bushings, or similar workpieces, in the ID for OD turning operations that have to be concentric.

Chuck a short piece of round stock and turn its OD to fit the ID of your bushings. This arbor now runs true.

Now file a flat on the arbor a few thousandths deeper than the diameter of a wire pin, welding rod, or small dowel—even paper-clip wire will do for small jobs. With the pin on the flat and the work slipped over both, spindle rotation rolls the pin off center and grips the ID. And as soon as the spindle stops, the work can be slipped right off. Accuracy is as constant as the IDs of individual workpieces, and work-changing is virtually instantaneous.

RICHARD F COOPER, *Plainsboro, NJ*

3.84 Stacked turning tools take heavy cut

Here's a method I have used to speed metal-removal rates by taking fewer passes at greater depths of cut, especially in interrupted cuts. In my experience, it's much easier on the toolbits and the machine, it results in a better surface finish on the work, and it can be used on both horizontal and vertical lathes.

One recent example was in machining some bridge expansion bearings, in which the maximum material removal was 1 1/2 in. deep. This was done with the illustrated tooling setup in three 1/2-in.-deep cuts—with each toolbit taking only a 1/8-in. depth and thus providing much softer entry and exit of the cut. The four bits in this case were 1/2-in. square, two above center and two below, with the bottom one projecting out the farthest and each successive one set back 1/8 in.

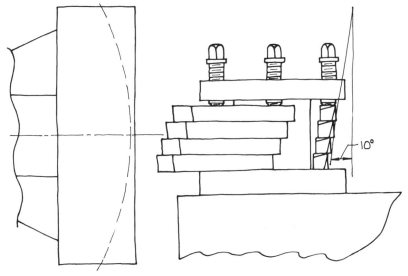

Viewed from the end of the tools, the bits were stacked at an angle of about 10°—or no less than the minimum angle that would match the helix of the cut. Special attention must also be paid to grinding the front clearance of the tools, especially the top one, to avoid heel drag.

D W O'LEARY, *Calgary, Alberta, Canada*

3.85 'Steady jaws' for long jobs

The required task was to bore and face a 4-in.-OD cylinder that was 12 1/2-in. long. No suitable steady rest was available for the Warner & Swasey T-5, which has only 2 1/2-in. bar capacity, that was our largest turret lathe. The workpiece overhang was horrendous!

We accomplished the job—meeting required dimensional and concentricity tolerances—as follows: 10-in.-long extensions were fabricated from 3/4- x 1-in. flat stock; these were bolted to soft jaws previously bored to size; and after clamping the work in the three-jaw chuck, the whole thing was firmly secured with hose clamps at the center and outboard end.

DONALD E CLAYPOOL, *Beaverton, Ore*

3.86 Stock-advance for hand screw-machine

The accompanying sketch shows a stock-advance mechanism I added to a hand screw-machine to help increase production of the machine.

The pivoting stock-advance lever and a 3/4-in. dia track are mounted on the machine bed or base, as illustrated.

When the collet closer is opened, the advance lever is pulled to the right. This initially actuates the "pincher" and then pulls on the sliding assembly to advance the stock. The adjusting screw is used to vary the amount of pre-travel

Turning

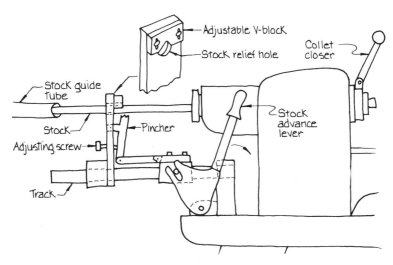

before the pincher engages the stock. And by using a hardened, adjustable V-block on the upright, the mechanism can be used for feeding a range of stock diameters—from 1/8 in. to 5/8 in. in our case.

The device works very well, and it has greatly increased production without going to the expense of an air feed.

MICHAEL SCHMIDT, *Owatonna, Minn*

3.87 Stop extenders speed multidiameter turning

Slip-on carriage-stop extensions provide a speedy, convenient, and accurate method for eliminating errors in turning workpieces with several diameters, shoulders, or grooves, and even internal jobs. The system is especially handy when used in conjunction with quick-change or indexing-turret tooling.

Just machine several sleeves to a slip fit over the stem of the lathe's carriage stop, and drill and tap a 6-32 hole near one end. Then, using rods that match the sleeve ID, grind them to the required lengths and insert them partially into the sleeves. Hold them with brass screws.

Use it best described with an example, such as the part shown in the upper left of the drawing and machined in the sequence indicated at the right:

1. Set the tool against the end of the work, install a 2 3/8-in. rod on the carriage-stop stem, adjust it against the carriage, and lock it in place. Now remove the rod and turn 2 3/8-in. length to the first diameter.

2. With no other changes, slip a 1/2-in. rod on the carriage-stop stem, and turn the second diameter.

3. Using a 1 3/8-in. rod, machine the outboard diameter.

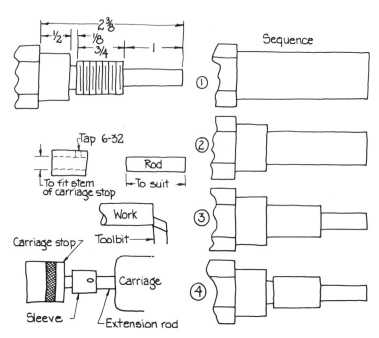

4. Reinstall the 1/2-in. rod and plunge-turn the 1/8-in.-wide thread relief.
The part is now ready for threading. Note that all tooling should be set prior to machining with the first length rod installed. For roughing cuts, just shorten the first length setting, and the others will automatically be short by the same amount.

BEN CRAIG, *Slatington, Pa*

3.88 Support your toolholder

Faster metal-removal with less chatter, especially with cutoff and form tools, is achieved by welding 3/8-in. Allen bolts to the bottom of lathe toolholders (as illustrated) for extra support. The long stud-coupling nuts on these bolts then provide for vertical adjustment and solid support. It works!

CARL FRANK, *Clark, NJ*

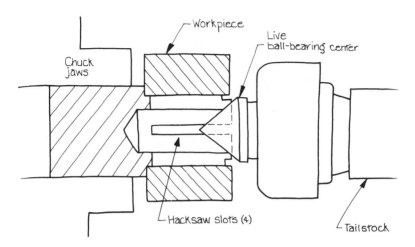

3.89 Tailstock center starts expanding mandrel

The drawing illustrates a simply made, accurate, internal collet or mandrel that can be loaded and unloaded without stopping the spindle and which only requires a live ball-bearing center in the tailstock for operation. It can be gripped in a three-jaw chuck or most other standard workholding devices.

The mandrel is made from barstock somewhat larger than the ID to be gripped. The center is drilled out to about two-thirds or three-quarters of the gripping ID and slightly deeper than the overall workpiece length. The gripping diameter should then be accurately turned to about 0.001 in. smaller than the workpiece ID, and the work-stopping shoulder should be faced. A 60° internal chamfer is bored at the end to mate with the tailstock live center, and two axial hacksaw cuts are made to provide four slots, after which the OD and chamfer are carefully deburred.

The workpiece is gripped when it's placed on the mandrel and the tailstock center is cranked in with some force. Accuracy depends on the care taken in machining the gripping diameter, the stop shoulder, and the 60° internal chamfer. Also the force applied on the tailstock should be approximately the same for each new part placed on the mandrel. Accuracy of the drilled hole is not critical, and a three-jaw chuck will give good accuracy if the mandrel is not removed or disturbed.

JAMES F MACHEN, *Toledo, Ohio*

3.90 Tailstock chucks support hollow cylinders

For the past several years we have been using a variety of standard lathe chucks set up so they can rotate freely when mounted on the tailstock. Fitted with special jaws, they are extremely useful and adaptable for providing

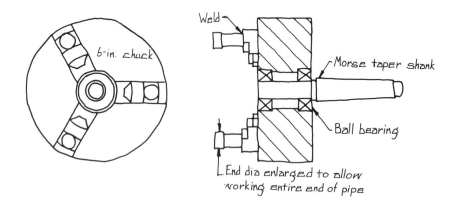

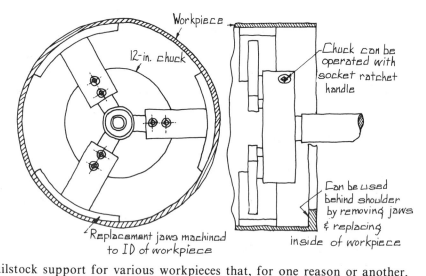

tailstock support for various workpieces that, for one reason or another, cannot be supported on a conventional tailstock center.

The first is a 6-in.-dia three-jaw self-centering type with inside/outside jaws welded to the chuck's original sliding jaws. Used mainly for pipe and tubes, the enlarged end diameter of these jaw extensions (see sketch) permit facing and even deburring of the ID when used internally. This chuck is mounted on ball bearings on a No. 3 Morse taper shank for direct insertion into the tailstock spindle of the machine.

A larger three-jaw chuck (12 in. dia) features replaceable top jaws, which can be tailored to suit a variety of work. The jaws shown in the second sketch are used inside 20-in.-dia pressure vessels for machining both the inside and outside surfaces of an internal flange. Although this chuck could also be mounted on a heavy taper shank, it's preferable for increased rigidity to mount it directly on a replacement tailstock spindle.

Also a small four-jaw independent chuck similarly mounted is very useful for holding irregularly shaped workpieces or shaft-type workpieces that have no center-holes.

R I CARTER, *McMinnville, Ore*

3.91 Tailstock tap-holder replaces third hand

The standard procedure for tapping work in an engine lathe—supporting the tap on the point of the tailstock center—often demands three hands. To simplify this operation, we built a sliding tapholder that fits the tailstock taper.

Using the taper shank of an old drill, we bored it to take a sliding bar, which was drilled axially to accept a tap shank. Setscrews against the tap's driving square prevent rotation and secure it in the bar. A dog-point setscrew in the Morse-taper section rides in a key slot cut the length of the sliding bar. Finally, a knurled collar is secured to the bar with another setscrew to permit the tap to be advanced or retracted manually.

CLINT MCLAUGHLIN, *Jamaica, NY*

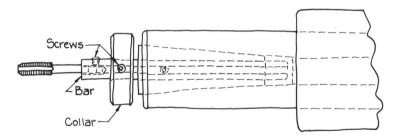

3.92 Taper attachment; offset equal steep taper

We had to turn some parts to a taper of 5.250 in. per ft—but our lathe's taper attachment was limited to 4 in. per ft.

At first, we set the compound slide at 12° 18′ and fed this manually for the taper. The hand feed, however, was tedious and did not produce a uniform finish. Furthermore, the taper was longer than the travel of the compound, necessitating a resetting of the carriage and contributing further to the lack of uniform finish.

Inasmuch as the parts were being turned on centers, we next tried setting the taper attachment at its maximum while simultaneously offsetting the tailstock center to produce the additional taper required. This permitted the full length

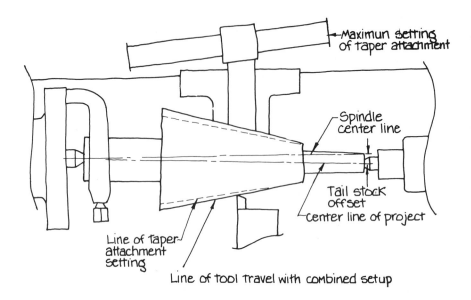

of the taper to be machined in a single setting and also allowed the use of power feed. Thus the operation was simplified, speeded, and now produced an acceptable surface finish.

The mathematics involved in finding the required tailstock offset is as follows: subtract the maximum taper-attachment setting from the required taper, divide this difference by 2, and multiply by 1/12 of the total length of the project. The result is the amount to offset the tailstock.

FRANK ACCURSO, Orinda, Calif

3.93 A tip for precision turning

When it's necessary to turn a very precise diameter on an engine lathe, try setting the compound rest at an angle of approximately 2° 52' to the centerline of the work. With that setting, an advance of 0.001 in. on the compound dial reduces the diameter of the work by 0.0001 in. (The sine of 2° 52' is 0.05001, hence the infeed on radius will be very close to 0.000,05 in. for every 0.001 in. of tool advance.) And it's an obvious advantage to be able to remove two or three "tenths" by turning the dial as many thousandths instead of trying to estimate on the cross-slide dial.

RAYMOND A MERKH SR, Acton, Mass

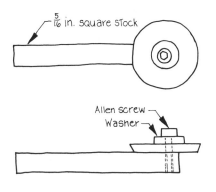
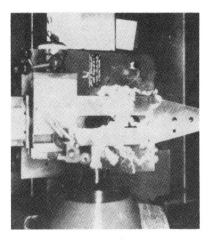

3.94 Tool cuts accurate radius in turning plastics

This lathe tool works well for turning accurate radii in plastic workpieces. It's made from 1/8-in. ground stock hardened and ground with a clearance angle, and it is simply screwed onto a piece of 5/16-in. square stock for a shank. Additional disks are easily made for the same shank, and should be marked with their radii.

The photo shows the tool used for cutting a 12.5-mm radius on a Delrin workpiece.

MARTIN BERMAN, *Brooklyn, NY*

3.95 Toolholder controls depth in centerdrilling

It is occasionally necessary to control the depth of center-drilled holes in a batch of parts being machined on an engine lathe. In the interest of efficiency,

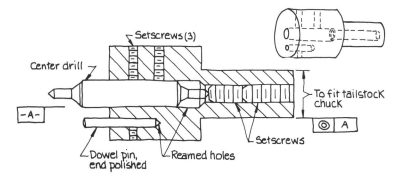

I have made several sizes of the tool shown in the sketch, which fit in the tailstock chuck of the lathe. They are especially useful when using the smaller (No. 11 and No. 12) "bellmouth" center-drills.

The center-drill is located axially for depth control by a pair of setscrews behind it, while a dowel offset radially provides the depth indication. The exposed end of the dowel is rounded and polished, and has not caused any scarring problems in the materials I typically work with, although more sensitive materials could be protected by using soft pins made of brass, plastic, or other stock. Both the center-drill and the dowel are secured with setscrews.

With minimal practice, it is possible to quickly and uniformly center-drill parts using the dowel as a stop—even when the work is obscured by coolant.

VAL KIEFER, *Seattle, Wash*

3.96 Tooling cuts NC cycle time

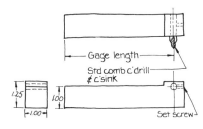

Many jobs that are run on NC or CNC turning machines start out with a center-drilling operation, followed by OD turning, facing, grooving, threading, etc. And many of these machines have two turrets—one primarily for external tools and the other for internal tools. The sequence of operations described above, however, requires running the end-working turret approximately 9 in. each time the center-drill is used, then running it back 9 in.

I have found considerable time savings by using the illustrated center-drill holder in the external-turning turret, which eliminates the need for 18 in. of traversing per workpiece. And, for many machines, use of a left-hand drill will also eliminate the need to reverse the spindle.

GEORGE ELLIS, *Alpharetta, Ga*

3.97 Turning small stock smaller? Support it

The accompanying sketch shows a specially made tool designed to facilitate the turning of small diameters on small rods for a greater axial length than

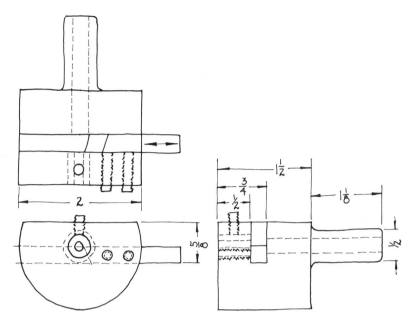

would be practical in a conventional manner. It does the job accurately, and in a single pass.

The tool is machined from a 2-in.-dia piece of cold-rolled steel. The 1/2-in. shank is turned first and a 5/16-in. hole is drilled through the entire length. then a 0.3755-in. hole is counterbored and reamed to accept standard bushings for whatever size stock is to be turned. Next, the top is milled off flat, and a 1/4-in.-wide slot is milled across to fit a 1/4-in.-square toolbit. And finally, the setscrew holes are drilled and tapped 10-32 to clamp the toolbit and secure the guide bushing.

Diameter of the turned portion of the rod is set by careful adjustment of the toolbit position.

WILLIAM HITCHEN, *Franklin Park, Ill*

3.98 A two-legged spider

We recently had a number of 8-in.-dia aluminum pipes that required facing off to a precise length. However, we had neither a large steady rest nor any suitable pipe spider to provide support at the tailstock end. A simple alternative was devised for the job:

We used a flat piece of 1/8-in. aluminum

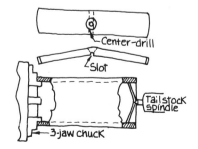

stock slightly longer than the ID of the pipe. The ends were then radiused to match the ID, and a slot about half as deep as the thickness of the material was cut across the center. The piece was then bent to a slight angle, as shown. With this wedged into the end of a pipe and center-drilled, tailstock pressure was easily sufficient to force the ends outward to provide adequate support for the facing operation.

HERMAN HAUSMAN, *Weymouth, Mass*

3.99 Use a boring head for off-center turning

One of our shop tricks for eccentric turning of small parts in a lathe is to use an adjustable milling-machine boring-head toolholder to hold the workpiece in the lathe spindle. It's generally a simple matter to devise the necessary workholding attachment itself, such as the threaded arbor shown in the photo, which is being used to turn a small plastic cam.

CARL FRANK, *Clark, NJ*

3.100 Winding special springs

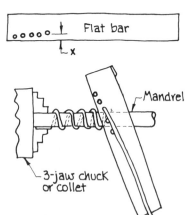

The need frequently arises for a special spring, and not only isn't a suitable one available in the shop, there isn't a commercial one that will do. Winding your own spring is usually the only answer.

This simple tool, used with a small bench lathe and an assortment of various size rods for mandrels, is all that's needed. Piano wire is usually the material for the spring itself.

The winding tool is made of a piece of 1/8-in. x 1-in. flat stock about 8 in. long. A series of 1/8-in. holes is drilled near one end at various distances from the edge, as at X in the drawing. This distance X determines the space between individual coils of the spring, and thus its pitch. It's a good idea also to lightly countersink both sides of the holes.

Diameter of the spring (after springback) is determined by trial-and-error winding of the desired wire size over a few mandrels.

To make a spring, the first few coils should be wound by hand before turning on the lathe spindle at a very slow speed. The winding tool should be held firmly, with the spring wire in the selected hole. It's also important to maintain even tension on the spring wire and not allow the end to run through the hole. After winding the spring, reverse the direction of the lathe spindle for a few revolutions to release tension before cutting off the wire. Either left or right hand springs can be wound, depending on direction of the lathe spindle's rotation.

WILLIAM HITCHEN, *Chicago, Ill*

3.101 Woodruff key clamps thin disks in ID

Especially for repetitive operations, clamping thin disks or large washers for facing or turning can be a problem. A split expansion arbor is generally satisfactory for small jobs with minimal torque, but for some jobs, the following technique may prove better.

Turn an arbor to provide a neat sliding fit in the disk, then mill a Woodruff-key slot, as shown, to accommodate a modified key. The key has a raised step on one end, and is flush with the arbor diameter at the other end.

The arbor is then mounted in a four-jaw chuck and indicated to run true. Three of the jaws hold the arbor in position, while the fourth one clamps down on the heel of the modified key, which will rock outward to clamp the work internally.

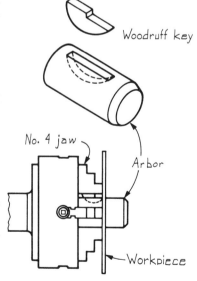

Only the fourth jaw must be actuated for every workpiece, and the system holds effectively under fairly heavy machining torque. For squareness, the workpiece is set against the jaws.

ALBERT T PIPPI, *Baltimore, Md*

4 Grinding and Finishing

4.01 Angle dresser is simple to make and to use

Here's an angle dresser for use on a surface grinder that can produce any desired angle on the grinding wheel when used with either angle blocks or a sine bar and the magnetic chuck. To cover the entire range of 0° to 90°, the

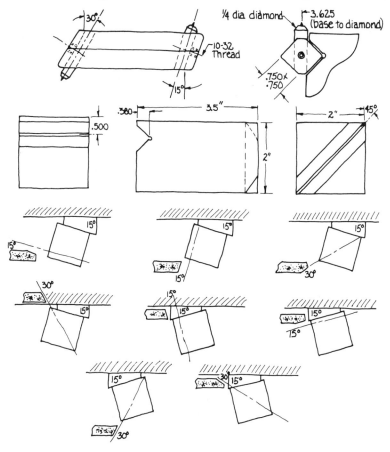

largest setting angle ever required is 22.5°; with any given angle block or sine-bar setting the simple dressing tool can be set up in eight different ways (see drawings, which illustrate use of a 15° block to add and subtract from the basic angles of the dresser itself).

The dresser is quite easy to make; its accuracy in use, however, is entirely dependent on the accuracy with which it was made, and so care is necessary. The original was made of 2-in.-square stock for the base and 3/4-in.-square stock for the slide. Dimensions are not particularly critical, except that the height from the base to the centerline of the diamond should be convenient; it's the angles that must be accurate and the surfaces that must be true.

The open slide may appear crude, but the intent was to create a design in which play would never become a factor and that could be cleaned in seconds. That is what this design offers. In addition, the sliding member can also be used directly on the magnetic chuck to hold a diamond for dressing the bottom or sides of the grinding wheel.

JIM MARTELLOTTI, *Irvine, Calif*

4.02 Avoid contact wear by diamond wheels

Whenever a diamond grinding wheel is mounted on a hub it should be dial-indicated to equalize any runout—especially on the periphery of a straight or plain wheel. The dual purpose is, first, to allow the wheel to run concentric with the spindle, and, second, to conserve diamond abrasive when the wheel is dressed.

A method I use to prevent excessive wear of the indicator's contact ball from the diamond is as follows: Place a 0.001-in. to 0.002-in.-thick feeler gage about 6 in. long in a small vise. Move the feeler gage to contact the wheel's periphery at or near its horizontal center line and flush with the wheel's surface. With the dial indicator mounted on a surface, bring the contact ball against the feeler gage at that point where the feeler gage is touching the wheel. Apply sufficient pressure to check the runout, and then turn the wheel. The feeler gage takes the wear, not the contact ball.

To finish up the job, tap the wheel to run true, tighten it down, dress it if necessary—and the job is done.

DAVID C RUUSKA, *Swartz Creek, Mich*

4.03 Ball centers for grinding tapers

Having to grind a 1/2-in./ft taper between centers with a toolpost grinder on a lathe, and not having any ball centers available, I simply made a pair from items that were on hand. Standard 60° centers just don't yield good enough results when the tailstock is offset by very much.

To fit the tailstock, I used the taper shank of a worn-out drill, facing off the end and carefully drilling a 60° centerhole in it. For the headstock, I just used a piece of barstock in a three-jaw chuck—again with a 60° center-hole in it. In both cases the center-hole should approximately match the center-holes in the workpiece. Then it was just a matter of offsetting the tailstock and mounting the workpiece with a bearing ball in the center-holes at each end.

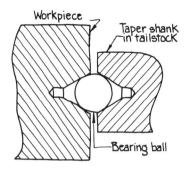

ROBERT MESSINA, *Blairsville, Pa*

4.04 Clamps flex to grip nonmagnetic parts

Two pieces of 3/32- x 3- x 4 1/2-in. ground stock, shaped as shown and hardened, can be used for holding round parts of nonmagnetic material on the magnetic chuck of a surface grinder. The pair of hold-downs are pressed inward against the circular piece as the magnet is turned on. This will put a downward force on the part, pressing it against the face of the magnetic chuck for grinding.

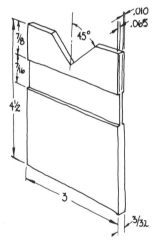

ALEXANDER WIENCKO, *North Quincy, Mass*

4.05 Controlled dump for deburring media

We use 1000-lb-capacity media tubs for loading deburring machines. These tubs were originally equipped with a single-hook lifting-plate, and a backbreaking manual assist was generally required to prevent the shifting load from dumping too abruptly. That problem has now been eliminated.

A controlled dump of media from tub to deburring machine was made possible by using an Owatonna Tool Co Load-Rotor gear-chain sling—with a somewhat modified rig. A new length of chain was added in the front—

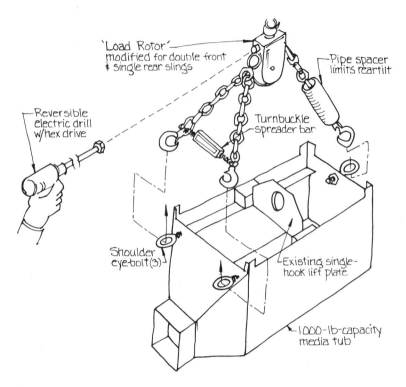

essentially "splitting" the two-chain sling into a three-chain sling. Three shoulder-eye-bolts were added to the media tub to suit this arrangement. A turnbuckle is used as an adjustable spreader on the front slings for ease of hookup. A length of pipe slipped over the rear chain limits the maximum tilt angle. And a reversible electric drill has all the power needed to drive the system.

GEDEON O TRIAS, *Rockford, Ill*

4.06 Damp vibration in grinding small punches

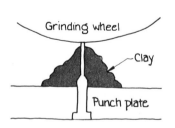

In the course of building, resharpening or repairing dies, it is frequently necessary to grind the tips of small perforating punches. Attempting this operation without supporting the part often results in a broken punch.

Here are two methods that will help when no other means is possible: (1) a mound of ordinary modeling clay packed around the

Grinding and Finishing 139

punch will damp out any vibrations set up by the grinding wheel, and (2) a rubber band wrapped around a group of perforators will put a helpful, light tension load at the tips.

WILLIAM HITCHEN, *Chicago, Ill*

4.07 Dresser puts 45° angle on grinding wheels

This special tool has been a great help to me in dressing grinding wheels accurately to 45° on a surface grinder with virtually no setup time at all. I use the 45° wheels for grinding die pins, mold pins, and die blocks.

The angle dresser is very simple to make and the sketch is pretty much self-explanatory.

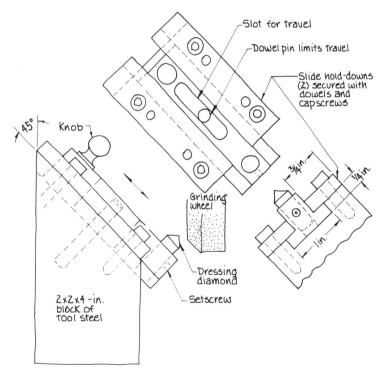

In use, the base block is simply set up on the magnetic chuck of the surface grinder, the wheel is brought down to proper position, and the diamond dresser is moved manually by means of the knob on the slide.

DAVID R LANDIS, *Chillicothe, Ohio*

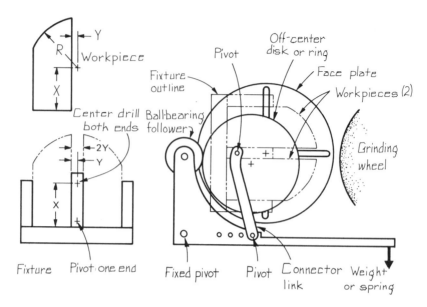

4.08 Fixture grinds a radiused workpiece

We had to grind a number of hardened tool steel cutting- and form-blocks shaped approximately as shown in the drawing. These could have been ground lengthwise on a surface grinder after dressing the wheel with a radius dresser, but this method would have resulted in a lot of wheel-dressing time (and re-dressing time) and in considerable wheel loss.

As an alternate, I built a fixture, as shown, with center-holes in both ends for use on a cylindrical grinder. Various surfaces of the fixture were ground to permit accurate miking of blade height without removing the parts from the fixture. The fixture is made to oscillate in the grinder by using an off-center ring mounted on the faceplate and the cam-and-connecting-rod arrangement illustrated.

This permitted us to use a regular grinding wheel dressed straight across.

DONALD D VAN HUIS, *Holland, Mich*

4.09 Fixture holds cutters for flattening shanks

This simple block-type fixture has proved to be very useful for grinding clamping flats on cutting-tool shanks. The steel block should be somewhat oversize to start with, so that the holes can be drilled and reamed without difficulty. The excess stock on the two outer surfaces should then be ground off on a surface grinder.

EMERSON L MOORE, JR, *Wichita, Kans*

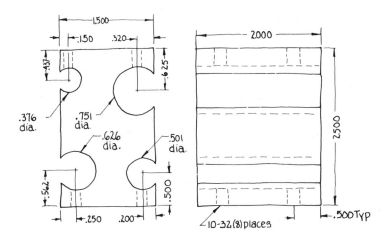

4.10 Getting stock square in a surface grinder

It's an easy task to grind two sides of a workpiece parallel in a surface grinder, but it's not always so easy to get the other two sides perpendicular to them. Here's a method that works:

First, grind two sides parallel. Then lay the part over and grind a flat-bottomed "dish" on one side, leaving a shallow rail at each extreme edge. Checking carefully with a precision square, grind off the high rail until the two rail surfaces are square with the sides. Next invert the piece and grind the "bottom" square with the sides. And, finally, re-invert the piece and grind off the rails and the complete "top" surface.

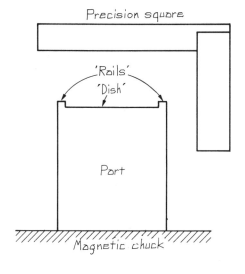

1. Grind sides parallel
2. Grind shallow 'dish' in top
3. Check with precision square, and grind high 'rail'
4. Invert part and grind bottom square with sides
5. Re-invert part and grind 'rails' off smooth

The technique works well, and it's a lot better than trying to use shims to achieve the same purpose.

JIM BAIAR, *Half Moon Rifle Shop, Columbia Falls, Mont*

4.11 Grinding crowns with a swinging fixture

If it's necessary to grind slightly crowned workpieces—either because the material is too hard for milling or because the required surface finish demands it—a large radius can be ground on a surface grinder by using the illustrated articulated fixture.

The workpiece is held on the swinging platform of this fixture by any suitable means. The platform is provided with trunnion pins, which extend through bushings in the sides of the fixture. The actuating arm is bored to fit the extending trunnion pin, and is then slotted and provided with a clamping

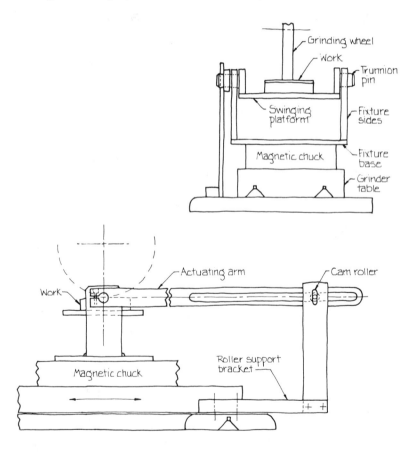

screw so the swinging platform can be positioned as desired. The actuating arm is slotted so it can slide on the cam roller as the table is traversed.

Supporting the roller is a bracket fastened to the cross-feed apron of the grinder. The vertical arm of the bracket is slotted so that the roller position can be adjusted up or down. If the roller is positioned on the horizontal center line of the slot and the trunnion pin, the swinging platform will remain parallel to the machine's table. If the roller is positioned above or below the center line, traversing the table will cause the actuating arm to swing in an arc, imparting a rocking motion to the swinging platform and causing the work to be ground in a large radius—convex if the roller is below the center line, concave if above.

FRANCIS J GRADY, *Reading, Pa*

4.12 Grinding small, compound angles

The drawing shows a useful method for grinding small, compound angles on a surface grinder when a compound sine plate isn't available, or when it's just too heavy and cumbersome to set up.

The drawing shows the workpiece and the magnetic parallel to be the same size. They needn't be, in which case the amount to shim the workpiece is proportional to the ratio of their respective sizes. For example, if the parallel is twice as long as the work, the shim under the parallel must be twice as thick as the amount of stock to be removed.

Note that the stock removed is the total of the thickness of shims under that corner. In case a magnetic parallel is not available, any accurate piece of flat stock can be used, in which case the workpiece should be "blocked in" to prevent it from flying off the machine.

MIKE CENTALA, *Ypsilanti, Mich*

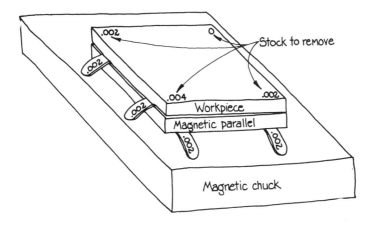

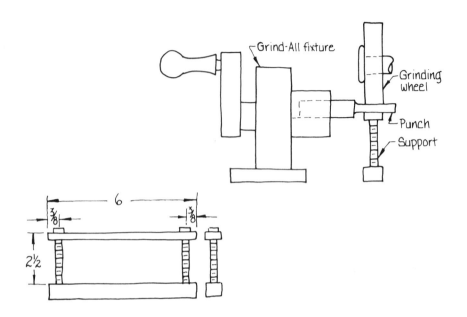

4.13 Heat-sink for grinding small punches

When grinding punches or locators in our Grind-All Grinding Fixture, the workpiece sometimes overheats or flexes, ruining both the dimension and the finish. It's particularly a problem when grinding the blank down to a small diameter for an inch or more back to the shoulder. The support shown in the drawing will prevent this problem.

The base is a piece of 5/8 x 5/8 x 6-in. cold-rolled steel, which is held by the grinder's magnetic chuck. The top bar is a piece of 1/4 x 5/8 x 6-in. copper, which not only provides support beneath the punch but also acts as a heat-sink. These two parts are joined by two 3/8-in. by 2 1/2-in.-long shoulder bolts with light springs around them made of 0.035-in. wire.

We find that the 2 1/2-in. shoulder bolts provide just the right height for use with our standard Harig Grind-All No. 1 fixture.

DON METZGER, *Louisville, Ky*

4.14 Improvise a stem-mounted diamond wheel

It happens that you've run out of stem-mounted diamond grinding wheels for use with a portable hand grinder and you can't wait for new ones before you have to break those carbide corners. It happens.

Just take a regular stem-mounted wheel, apply a bit of coarse diamond paste to it by pressing the paste into the porous grit surface of the wheel with your fingers. and then try grinding that carbide section with it. It works.

BEN SCHNEIDER, *West Orange, NJ*

4.15 Master tool block for surface grinder setup

This tumble-block-type tool-grinding fixture greatly facilitates setups for generating the precise compound angles required on threading, grooving, and parting tools. With its built-in angles, its setup is simply a matter of clamping the square toolbit in place with the setscrews and placing the steel block on the magnetic chuck of a surface grinder.

GENE BROLUND, *Mentor, Ohio*

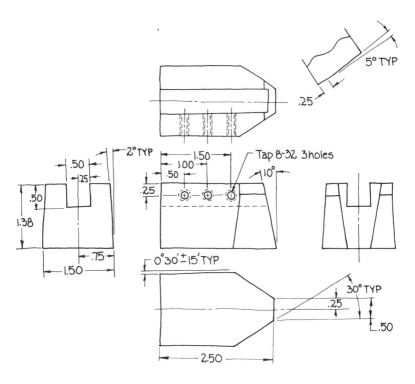

4.16 Quick-change stop for magnetic chucks

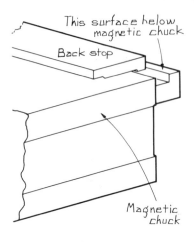

When it's necessary to work over the rear edge of a magnetic chuck, it is always a tedious job to remove the backstop and then replace it for further use. The screws and screw holes are full of grit, and then the backstop has to be reground after it is replaced because it never seems to go back in place exactly the way it was before.

The backstop sketched here, however, is a lot simpler and quicker; you just lift it off and then lay it back down. You do have to wipe the groove out to remove grit, but that's a lot easier than cleaning out a series of holes. Note that the rear component, which is bolted to the magnetic chuck, never has to be removed because its top surface is lower than the top surface of the chuck. The removable backstop is hardened and ground.

BALLARD E LONG JR, *Oak Ridge, Tenn*

4.17 Simple fixture restores spade-drill

Occasionally you have to grind the OD and relief of a spade-drill blade to restore it to working condition, which is often preferable to the alternate method of grinding back the cutting edges. The problem with the latter method is that it usually seems to require too much grinding and thus shortens blade life.

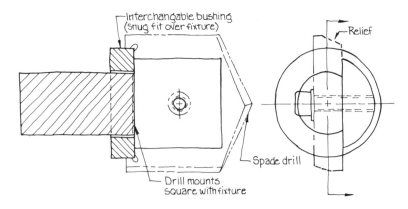

The method I use requires a surface grinder, a Hardinge head or a Harig Grind-All, and the simply made fixture illustrated here.

Shank of the fixture is turned to fit a standard 5c collet; a set of snug-fitting interchangeable bushings will position the "ears" of various-size spade drills; and shims or spacers on the fixture's flat will center the thinner blades. To grind the relief, use the divisions on the grinding tool that holds the fixture.

MATTHEW FORTUNATO, *Colonia, NJ*

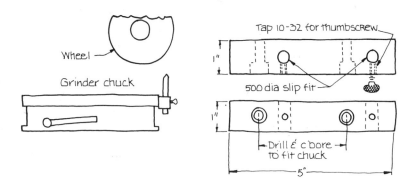

4.18 Mount holds dressing tool out of way

By mounting this easily made diamond holder in existing holes on the end of the magnetic chuck, it's possible to dress and redress the wheel of a surface grinder without removing workpieces from the chuck.

Also, by using a diamond dressing tool with a long shank, the job can be done without lowering the wheel. A thumbscrew clamps the dressing-tool shank, allowing adjustment and also permitting the diamond to be lowered below the height of the work during grinding operations.

STEPHEN G ROBY, *Flint Tool & Machine Co, Alexandria, Ind*

4.19 Soft arbor gives firm ID grip, no distortion

The illustrated arbor, which can be used either for turning or grinding, is very useful for gripping fragile workpieces on the ID for machining without the distortion frequently encountered in more-conventional clamping. Depending on the precision with which the arbor itself and the ring are machined (and the tolerance of the existing workpiece bores), both radial and axial accuracy and repeatability can be very good, and the ability to grip the work securely and drive it are quite satisfactory.

Tightening the capscrew simply drives the ring inward to compress the

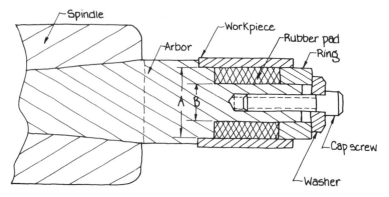

rubber sleeve, which expands to grip the work, easily preventing it from slipping on the arbor during machining.

Diameter A is, of course, determined by the workpiece, and diameter B by whatever rubber sleeve may be available. Ready-sized sleeves are available in most cases, but can also be machined to suit without too much difficulty.

FRED STAUDENMAIER, *Bedford, Quebec, Canada*

4.20 Support that grinding job

We had to make straight-flute drills out of hardened 5/16-in. drill blanks for screw-machine starting drills. I could only grab the 6-in.-long blanks by about 3/4 in. at the shank end to grind the required 3-in. flute length before the wheel would start grinding the work-head. Naturally with this small-diameter blank hanging out that far it was impossible to grind because of the chatter.

I made a small screw-clamp, as illustrated, out of 11/16-in. hex brass stock to fit a Craftsman magnetic base. This allowed the blank to be held securely, and it virtually eliminated the vibration, permitting the drills to be made without further problems.

ROB MOREL, *Walla Walla, Wash*

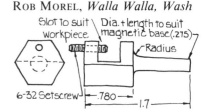

4.21 Surface grinder shapes form-tools easily

A simple steel block approximately 2 in. x 2 in. x 3 in. high with a 1/2-in. angled slot and two tapped holes for setscrews makes a very useful fixture for grinding form-tools on a surface grinder.

The angle of the slot, in which 1/4-in. to 1/2-in. toolbit blanks can be clamped, determines the end relief of the tool. And setting up the block with a protractor against the side of the magnetic chuck (see sketch) determines side clearance.

MARTIN BERMAN, *Brooklyn, NY*

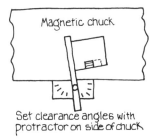

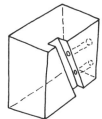

4.22 Surface grinding fine punches

We had a large progressive die to make that called for a dozen 0.020-in. piercing holes. To produce the punches quickly on a surface grinder, we made this simple fixture.

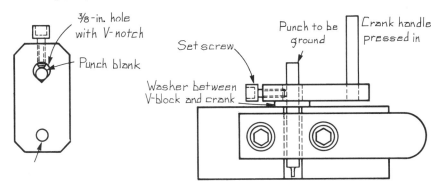

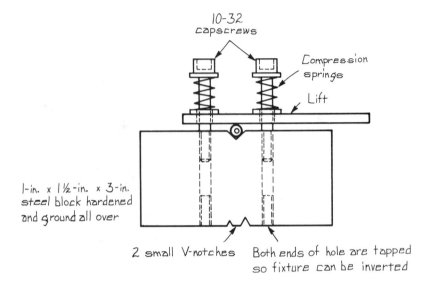

A block of oil-hardening tool steel (1 in. x 1 1/2 in. x 3 in., but dimensions are not critical) was drilled through as shown and tapped 10-32 for about ten threads from the top and from the bottom, which allows the fixture to be used inverted. The block was then hardened and ground square on all six sides, and three V-notches were ground to various depths to accommodate different diameters of drill-rod punch blanks from 1/16 in. to 3/8 in.

The spring-loaded pressure pad (left long on one end for easy insertion and removal of the punch blanks) and the crank throw were made up of other bits of tool steel. The latter component had a dowel pin crank handle pressed in at one end and a 3/8-in. hole with a V-bottom cut in the other to accept a punch blank that is clamped in place with a capscrew.

In use, the fixture is placed against the back parallel of the surface grinder's magnetic chuck, the grinding wheel is carefully centered over the punch blank, and the crank is slowly turned to grind the punch to proper diameter. We found it convenient to grind down both ends of a double-length blank, and then cut it in half to produce two punches.

RUDOLPH K SCHMITT, *Whitestone, NY*

4.23 Switch to surface grinder speeds output

A cylindrical grinder was used to grind all three of the surfaces indicated in the upper drawing. Surfaces B and C were ground together, then surface A was ground in a different setup on the same machine. An

Grinding and Finishing

increase in production requirements, however, exceeded our capacity on the single machine we had, and investment in another cylindrical grinder was not justified for this operation. We did have some spare time on a surface grinder however, so it was suggested that the process be modified to use that machine.

In the revised process, surfaces A and B are ground together on the cylindrical grinder, and a magnetic transfer block is used on the magnetic chuck of the surface grinder to permit surface C to be ground parallel to surface A. Not only were machine capacities better balanced, but also the cost of grinding the part, a gear, was drastically reduced by using a number of transfer blocks and grinding a gang of parts at one time on the surface grinder.

BIPIN C BHOGAL, Poona, India

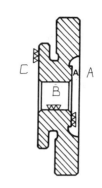
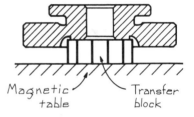

4.24 Upside-down grinding big jobs

Putting oversize jobs on undersize machines is a setup problem that occurs over and over in any shop. At times all operators of surface grinders have run

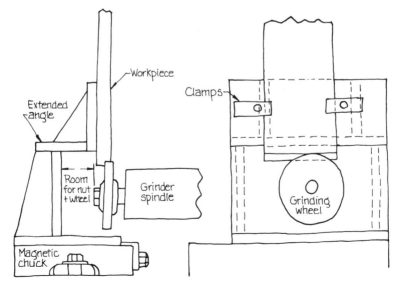

into the problem of having to grind the end of a workpiece that's too tall when upended under the wheel or too long to be supported on the grinding chuck.

So why put the work under the wheel? To overcome this frequent situation, I constructed, from flat ground steel, an extended "angle iron" that allows the wheel to be used under the workpiece. This allows grinding virtually any size workpiece that isn't obstructed by overhead clearance.

Good accuracy was obtained by using stress-relieved steel, and by screwing and doweling it together. After assembly, the angle fixture was squared and precision ground. Holes are drilled and tapped in the fixture for strap clamps to hold the work.

The wheel guard must be reversed, of course, to go under the wheel—and it's important for the grinder operator to remember that downfeed is now up.

WINTHROP LORING, *Rochester, NH*

4.25 Vertical spindle for surface grinders

You can grind both sides of a dovetail in a single, simple setup on any ordinary horizontal-spindle surface grinder—and facilitate some other jobs as well—if you add a vertical spindle to the machine.

Just make a right-angle bracket as shown in the sketch. One end fits the outer sleeve of the surface grinder's wheel spindle, and the other accepts a small, high-powered, electric hand grinder. These holes should be carefully

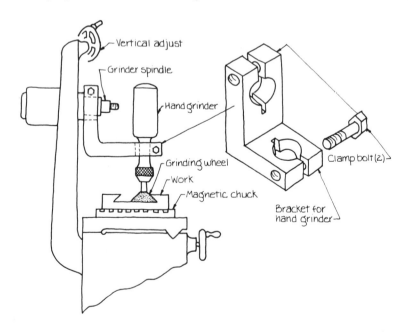

bored at right angles, then be slotted, and finally be drilled and tapped for the clamp bolts. And in setting up the vertical spindle, care should be taken to get it truly perpendicular to the table as viewed from in front of the machine.

RICHARD L RODGERS, *Warren, Pa*

4.26 Wear it down—but evenly

Most machinists and toolmakers tend to use the center of a magnetic chuck every time. This tends to wear the surface in only one section. It's better practice to place jobs at random locations on the magnetic chuck in order to both reduce the degree of wear and distribute the wear evenly over the entire area. The same practice also applies to other machines such as jig borers and milling machines.

PAUL CAPUTI, *Bloomfield, NJ*

5 Cutoff and Sawing

5.01 Bandsaw cuts plastic foams

Our shop occasionally has to fabricate parts from various types of foamed plastics—both rigid and soft. Cutting these parts with a bandsaw resulted in a flurry of dust and an unacceptable cut.

To do the job properly, we made a knife-edge blade from a discarded bandsaw blade. Using a bench grinder, we first ground off the teeth. Then a fast pass across the grinder was used on both sides of the band to give it a fine, serrated edge.

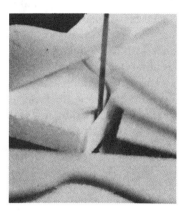

This blade easily permits us to cut and slice the problem materials, including Styrofoam (which we use for packaging delicate instruments for shipment).

CARL S FRANK, *Clark, NJ*

5.02 Carbide guides for bandsaw blades

The 1/2-in.-sq blade guides on our much-used Rockwell Delta 14-in. vertical bandsaw suffered considerable wear so that they were no longer effective in holding the blade accurately for straight or contour cutting. I have found that used tungsten carbide inserts can be silver-soldered to the ends of the solid guides, and that they practically never need to be adjusted and don't wear out. I used Valenite SNC-431 inserts in VC55 grade, but virtually any 1/2-in. square insert designed for negative rake use (0° clearance on the sides) would also do the job. If the inserts are coated with ceramic or titanium nitride, it will first be necessary to grind off the coating so that it can be silver-soldered.

STEPHEN G PEETZ, *Wahoo, Neb*

5.03 Contour 'burning' on an NC mill

You can convert a numerically controlled milling machine into an NC flame-cutter simply by mounting a manual torch—aimed downward, of course—on the end of a rectangular steel bar and clamping this "outrigger" onto the machine table. Then make a grid table to fit an adjustable die cart. To get the correct height for cutting, raise or lower this table, or make the vertical adjustment by raising or lowering the mill's knee.

DON WIEDERHOLD, Kansas City, Mo

5.04 Cutting heavy gaskets

We occasionally have to cut large gaskets from heavy, stiff, relatively brittle sheet gasket material, which tears too easily to cut with shears and is very laborious to cut with a knife. We've found that this material can be cut quite successfully by laying it up on a piece of plywood and using an inexpensive, hobby-type saber saw.

The secret is in the blade: a short one ground with a chisel-like edge on the bottom. Blade length is selected so that the cutting edge just goes through the gasket material and a few thousandths of an inch into the plywood backup sheet on the bottom of its stroke. This will easily follow either a straight or a curved line, and it leaves a clean, smoothly-cut edge on the gasket.

S T CRAWFORD JR, Ashland, Ky

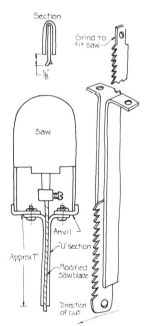

5.05 Cutting 6-in. foam plastic

We occasionally have to cut foam plastic into intricate shapes for padding in tool cases and similar applications. The material, which we use in thicknesses up to 6 in., is easily cut with the simple attachment we made to fit on an inexpensive home-workshop-type of saber saw.

The U-section extension is made to 0.020-in. sheet steel bent to fit a hacksaw blade and with two drilled and bent tabs at the top end for mounting. Width should be such that the hacksaw-blade's teeth project about 1/8 in. out of the front.

The upper end of the blade is ground to fit the saber saw. We use a blade with 18 teeth per inch arranged so it cuts on the down stroke, although it would probably work equally as well cutting on the up stroke.

D THOMPSON, *Redmond, Wash*

5.06 Drill a 'drain hole' to clear hole-saw chips

The best way I know of to eliminate chip-disposal problems and the clogging of teeth in hole-sawing operations is simply to provide a "drain." Merely drill a through-hole tangent to the inside of the periphery of the desired hole to allow the chips to fall through. Approximately a 3/8-in. to 1/2-in. hole is recommended.

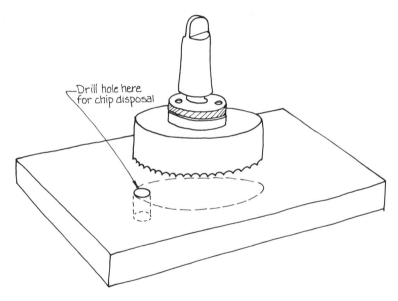

The simple technique saves a lot of time and hassle whenever the thickness of the material to be hole-sawed is greater than the depth of the saw's teeth. It eliminates blowing chips all over the shop with an air nozzle, woodpeckering, and brushing chips out of the narrow kerf. It's exceptionally helpful on thick plate, say 1 in. to 2 in. thick.

JERRY FORCIER, *Petaluma, Calif*

5.07 'Drop-leaf' table facilitates torch jobs

At our plant we frequently are called upon to cut holes in dished steel tank heads with an oxyacetylene burning machine. Holes may be required in the

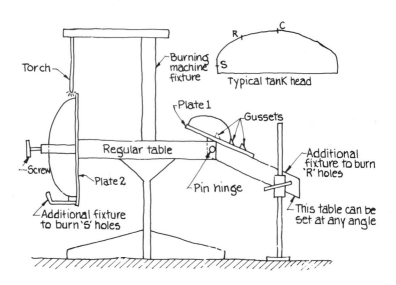

center of a head ("C" in the sketch), eccentrically somewhere else on the radiused end (R), or in the straight cylindrical flange (S). The center holes pose no problems, and have always been burned with the work held on the machine's regular table. Holes in the other two locations, however, had to be done manually.

The existing table and fixtures were modified, as shown in the sketch, to eliminate this problem.

To burn eccentric holes in the R surface, the pivoted table at the right was added, which can be adjusted to any desired angle. A plate (1) was welded on, to which were welded a number of small, triangular "gussets" to support various-sized heads. The head is simply placed on this table, and the table is lowered until the surface where the hole is to be burned is level—an adjustment that needs to be made only once if several identical heads are to be produced.

To burn holes in the flange (S) surface, plate (2) was welded to the regular table on the left side. This provides simple support and clamping arrangements, as shown.

KISHAN BAGADIA, *Waterloo, Iowa*

5.08 Jig aids bandsawing many short keys

Sometimes in assembly work we find occasion to make a quantity of short keys, usually of 1/8-, 3/16-, or 1/4-in. square keystock. Using a hacksaw would be impractical, and using a bandsaw could be hard on the hands. So we avoid the problem by making a simple jig.

Just take a rectangular block of steel or aluminum, and drill a hole—or several different size holes—just large enough to accept the keystock and to the same depth as the required length of the keys. And if different lengths are sometimes required, a tapped hole on the opposite side will provide the means for adjusting the lengths.

JOHN Q TILLMAN, *Akron, Ohio*

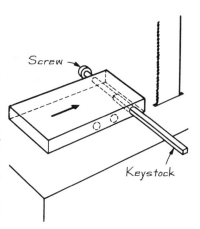

5.09 Make die-set casters

Sawing heavy die-set components on a vertical bandsaw can become almost impossible when the weight of the parts gets so great that they're difficult to push through the saw. When this problem arises, we use what I call "die-set casters," as shown in the sketch.

They're simply squares of aluminum plate drilled with a pattern of holes for steel bearing balls, which are held in place by staking on both sides with a modified cold chisel. The casters can be used with any heavy, flat material as an aid in sawing.

JOHN J LITGEN, *Carpentersville, Ill*

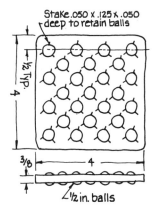

5.10 OD pilot for hole saw improves accuracy

When cutting holes in sheet metal or plate with a hole saw, the saw may tend to "walk" away and even break the pilot drill. This is especially true when using a portable drill. The following procedure helps prevent this:

Using a hole saw of the diameter required for the job, cut a hole in a hardwood board. Now center the hole on the proper workpiece location and

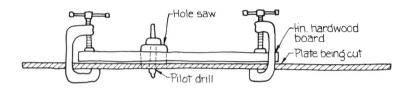

clamp the board in place by any suitable means. And finally saw the hole through the sheet or plate. The method works equally well with either a drillpress or a portable hand drill. It not only prevents the hole saw from walking, but also provides a reservoir for cutting oil.

JAMES E HARRIS, *Winston Salem, NC*

5.11 Sawing radiator hose

Faced with cutting a rather large quanity of rubber radiator hoses on a one-time production basis, we tried a number of methods without much success. Then a bright idea! Take a standard 10-in.-dia circular ripsaw blade, sharpen the teeth like knives, and run it backwards.

It worked like a charm. But be sure to run the blade backward or the teeth will grab the hose instead of slicing it.

W E TRITZ, *Waukesha, Wis*

5.12 Shear cuts wires to length at 100 rpm

In the absence of any press suitable for cutting short lengths of wire up to about 0.100 in. diameter, the job can be done in a drillpress with the tooling illustrated here. Cutting lengths of 0.080-in.-dia wire in a 3/4-hp machine, we hand feed wire with the spindle running 100 rpm.

Essentially, the device consists of a cylindrical die fixed to a base, and a "punch" powered by the machine spindle so that it rotates within the die. Cutoff length of the wires is fixed by either the upper stop or the lower stop, and changing to totally different lengths just requires simple replacement with stops of different diameter.

In making the cylindrical die, it's a good idea to make many holes of each wire-diameter that is likely to be cut. As the die wears, you can then just shift the operation to another hole of the same diameter. And when the rotating punch is worn or damaged at any particular point, it can be raised or lowered

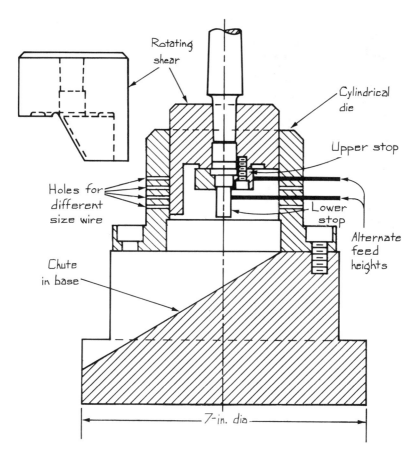

by adjusting the machine's quill height. Finally, the punch can be resharpened many times.

JOUKO I LEHTONEN, *Billnas, Finland*

5.13 A tip for hacksawing

Here's an easy way to saw a piece of round steel barstock with a hand hacksaw: Rock the saw across the piece instead of cutting it flat. This will make your cut truer, and will also prolong the life of the blade by helping to clear chips from the blade's teeth.

MARTIN J MACKEY, *Parma, Ohio*

5.14 Welding bandsaw blades

To assure a perfect match of the ends for welding bandsaw blades, I hold the blade ends together with the teeth on opposite sides and grind them both at the same time. Any deviation from squareness is thus exactly compensated when the ends are reversed, and results in good welds every time. The photos show a job that was purposely misaligned while grinding for illustrative purposes. And they match.

CARL ROSSMANN, *Euclid, Ohio*

6 Threading

6.01 Adapter sleeve keeps tap in line

Small taps are fragile things, too easily broken—especially if the hole to be threaded is in an odd position or at an angle other than perpendicular to the work's surface.

If you can drill the hole, however, this simple adapter sleeve will ensure that you can keep the tap perfectly aligned and thus prevent tap breakage due to misalignment. It's just a short length of barstock turned down at one end to fit the chuck with which the tap-drill was held and with the other end drilled out about 2 in. deep for a slip fit on a 3/8-in. or 1/2-in. shank. This latter shank mounts a smaller chuck for holding the tap, and it can also be cross-drilled for a short bar so the tap can be rotated by hand into the work.

The method can be used in a drill-press, milling machine, or lathe. Just make sure the workpiece doesn't get shifted between the drilling operation and tapping.

RON STANWICK, *Englishtown, NJ*

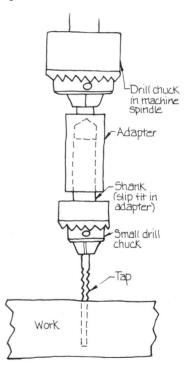

6.02 Boring/threading bar does double duty

Good planning can enhance the profitability of even the most routine machining jobs. We had a quantity of cast-iron hubs that required boring and threading. Because no turret lathe was open, we set up the job in an engine lathe and made a double-tooled boring bar, as shown, to permit the job to be done in a single chucking without changing tools.

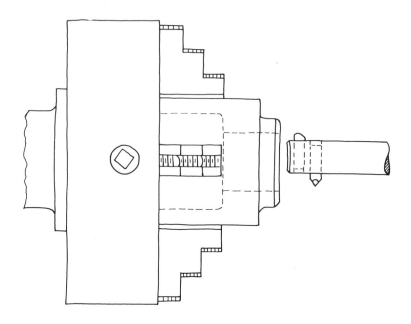

The rear tool faces down and is used to bore the hole to size. The front tool is used for threading and is set up conventionally. The only adjustment between the two operations is to reduce spindle speed and shift the feed for threading.

CLINT MCLAUGHLIN, *Jamaica, NY*

6.03 Chucking taps—securely

Conclusive evidence that drillpresses are often used for tapping holes is found in the number of drill-chuck keys with bent handles that result from the use of a pipe extension for tightening the chuck so the tap wouldn't slip. But it only takes about five minutes to put a tap in an indexing fixture and grind three small flats on the shank. These flats will match the jaws of any drill chuck, and prevent the tap from turning even though only hand tightening force is used.

GENE MAIER, *Jackson, Miss*

6.04 Double-check thread pitch

Here's a practice I follow, which has saved me from producing a lot of scrapped parts. Before cutting threads on a lathe, I first divide 1 in. by the number of threads to find the pitch—or the distance the carriage must travel for one revolution of the spindle. For example, to produce a 1/4-28 thread, the carriage must travel 0.0357 in. per revolution.

Next I engage the carriage and remove the backlash, and then zero the travel dial and determine how far the carriage moves when I hand-turn the spindle one revolution. If it isn't very close to the figure I calculated, then I know I'm set up for the wrong thread.

Once I found the machine was set up for metric operation; if I hadn't double-checked myself, I wouldn't have noticed it in time.

MIKE BOWLING, *North Vernon, Ind*

6.05 Fixture enables milling long-pitch helixes

It's a rare lathe that will cut threads of greater pitch or lead than 1/4 in. To produce an even longer lead (for the feedscrew of a plastics extruder), I recently used a Bridgeport mill with the special fixture shown in the sketch.

A shaft is mounted in two pillow blocks on the table. One end of this shaft drives the workpiece, as shown, while the other is driven through a non-slip drive (Sprocket-and-chain, timing belt, or even gears will work) by the table's leadscrew. The handcrank, graduated bezel, and lockscrew are removed from

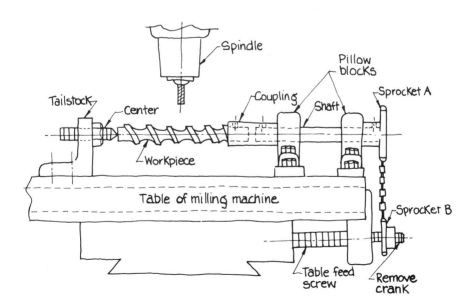

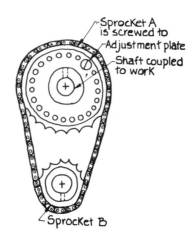

the leadscrew and replaced by sprocket B. Ratio of the number of teeth on sprocket A to those on sprocket B ($N_{AA}|A / N_{AB}|A$) is equal to the desired lead on the workpiece divided by the pitch of the table leadscrew. (Leadscrew pitch on a Bridgeport is 0.200 in.)

With the setup illustrated, in which both sprockets rotate in the same direction, a right-hand helix will be cut. If a left-hand helix is desired, it's necessary to introduce a gear or use a crossed (or twisted) timing-belt arrangement. The later scheme, however, is less satisfactory than gearing.

If extremely wide flights are desired, requiring multiple passes of the end-mill, or a multiple-start thread, a "phase adjuster" can be used (see smaller sketch). This involves a drive plate mounted on the fixture shaft to permit the sprocket position to be radially relocated. This works better than trying to adjust the radial position of the workpiece in the coupling (or chuck).

JAMES F MACHEN, *Toledo, Ohio*

6.06 Forked fork makes speedy diestock spinner

Chucking a tap in a variable-speed, reversible electric drill to "chase" rusted or damaged threads in a threaded hole or nut is common enough machine shop practice. But doing the same job on bolts or threaded rods is typically a manual task: running the die on and off by hand, with the die held in a diestock.

You can do something about that, however, as I did recently when I had a bucketful of rusty 5/16-in. by 5-in.-long carriage bolts to recondition. I made up a simple "spinner" for turning the diestock on and off with a reversible, variable-speed drill.

First I bent a "U" of 1/2-in. steel rod with about 2 1/2-in. spacing between legs and approximately 5-in. length. I then welded a short length of 1/2-in. rod at the center of the base of the U for chucking and two short pieces of 1/8-in. by 1/2-in. flat stock at the end of each leg, making sure that the latter pieces were properly spaced and aligned to fit the handles of my diestock.

Procedure is simple and obvious. Just clamp the work in a vise, add a dab of cutting oil, start the die by hand, then—with the fork already chucked—just run the diestock on, then off with your drill. It only takes seconds for each bolt. And if your drill has the power, it's even possible to cut new threads or extend the threaded portion of bolts.

ANDREW VENA, *Philadelphia, Pa*

6.07 For non-slip tap-chucking

The illustrated tap-holder is simply made and eliminates the alternate evils of either overtightening a drillpress chuck or having the tap slip. The flats on the shank prevent the holder from rotating in the chuck, and the snug hole and setscrew arrangement at the bottom effectively align and secure the tap.

DAVID A SAAM, *Milford, NJ*

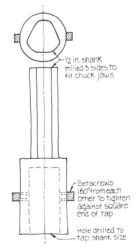

6.08 'Gage' sets thread-tool height, angle

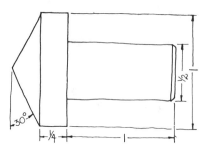

This little gadget speeds tool-setting for threading in a lathe by allowing you to set both tip height and angle simultaneously. Just chuck the setter in the tailstock, put the tip of the threading tool at the point of the cone, and line up the tool's edge with the taper.

One modification that might be preferred by some would be to turn the gadget's shank to a taper that fits the tailstock instead of the straight shank illustrated.

H H JACOBS, *Franklin, Wis*

6.09 Hand-tapping station added to drillpress

Drilling and then tapping is a way of life in the tool shop, and to facilitate this frequent sequence we have added hand-tapping stations to some of our drillpresses.

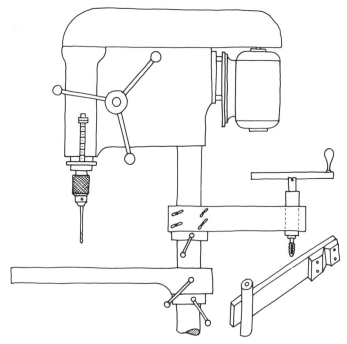

Threading

The unit consists of an extension bar with a clamping ring mounted below it on the column. It's also desirable to mount a clamp ring below the table as well. These rings are available from the drillpress manufacturer.

A plate is added to the inside of the extension bar so a V can be milled in it to keep the bar square with the column. A tube is welded to the outer end of the bar, as shown. Center-to-center distance from the column to the tube should be the same as to the drill spindle. The extension bar is clamped to the column with U-bolts and wing-nuts.

Taps are held in individual hexagon nose pieces, which slip into the bar. These nose pieces and the bar and crank handle are available from manufacturers of hand-tapping machines.

To use the setup, the work is clamped to the drillpress table, and the first hole is drilled. The table is then unclamped and swung around to the back, the tap is lowered, the table is locked into alignment, and the hole is tapped. Then the work is unclamped and repositioned for the next operation.

Although the drawing shows the tapping unit at the rear, we usually find it most convenient at the side of the machine.

CLINT MCLAUGHLIN, *Jamaica, NY*

6.10 Holder aligns die for threading on a lathe

Whenever it's necessary to cut a small thread with a die in a lathe, keeping the die straight for starting can pose a problem. When the thread must be a couple of inches long, the result may be an undesirable wobble. This gadget avoids the problem, and it's simple to make.

To make the first part, I simply used the taper section of a broken drill and welded a 6-in. length of 1/2-in. drill rod to it, being careful to keep it straight and concentric. The second part is a 6 1/2-in. length of 1 1/2-in. brass barstock with the ID bored out to a slip-fit on the 1/2-in. drill rod, one end bored to accept 1-in.-dia threading dies, and the OD knurled for a secure grip. Appropriate holes are also drilled and tapped 90° apart for setscrews to hold the die.

Use of the tool for threading in a lathe is obvious.

ELBERT A HAZZARD, *Largo, Fla*

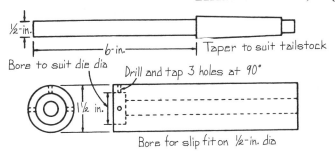

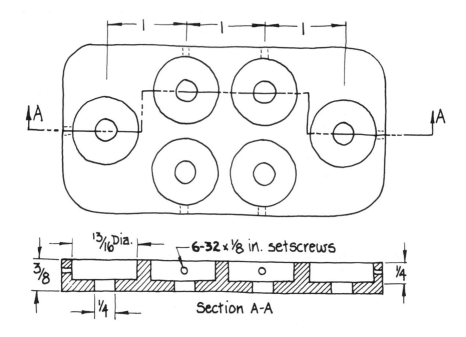

6.11 Holder stores six threading-dies

Whenever it was necessary to make up some special screws or clean up some fine threads that had been burred, I always had to search for the proper die or a die-holder. That hasn't been a problem ever since I made up the simple gadget shown in the sketch—a multiple die-holder that not only keeps a half dozen dies in one place but that also serves as a die-holder or diestock. It even saves the time previously needed for clamping the die into the holder. Machining the die-holder from aluminum was easy.

JOHN URAM, Cohoes, NY

6.12 Mental tap-drill chart

One of the easiest ways to determine tap-drill size for any 60° thread is to subtract the pitch from the major diameter and select the closest drill size.
   Some examples:
   For 1/4-20; 0.250 − 0.050 = 0.200 in., use a No. 7 or 13/64-in. drill.
   For 1/2-13; 0.500 − 0.077 = 0.423 in., use a 27/64-in. drill (0.422 in.).
   For a metric M 14 x 2; 14 − 2 = 12 mm.
   The method can be used for any 60° thread, no matter what size.

JOHN E WHITTLES, Roseville, Minn

6.13 Notched tap saves work

Everybody knows how easy it is to break a small tap. And it can almost be a disaster when one of these small broken taps cannot be removed from a complicated or expensive workpiece. I think most machinists would find it preferable for the tap to break at some location where it can still be gripped and backed out of the work.

To make this happen, I grind a circumferential notch in the tap shank near the squared driving end of the tap. Depth of the notch is such that the diameter at its bottom is slightly less than the minor diameter of the tap. This notch then becomes the tool's "weak link," and that's where it will break.

The grinding job is done on a surface grinder with a hard, fine wheel dressed down to about 1/64-in. width at the periphery. The tap is held in a collet in a fixture such as a Hardinge Dividing Head, and is rotated under the wheel.

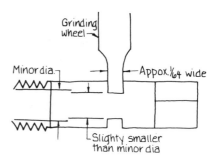

WILLIAM HITCHEN, *Chicago, Ill*

6.14 O-rings float diehead

Here's a simple diehead that floats for self-alignment with the part to be threaded. The dieholder is a loose fit—about 1/32 in. clearance on each side—in its sleeve, and is held centrally but flexibly by two O-rings in internal grooves. At the rear is a thick rubber washer (behind a separating steel washer) for axial pressure with a bit of float as well. The assembly is held together by a pair of pins pressed tight in the sleeve, but loose in the dieholder. Finally, a pipe fitting at the extreme rear end allows for a blast of air to clean out the chips after each threading operation.

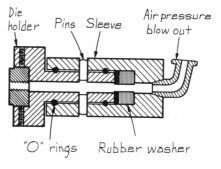

ROBERT J PHILLIP, *Oshkosh, Wis*

6.15 Recycled chasers make threading bits

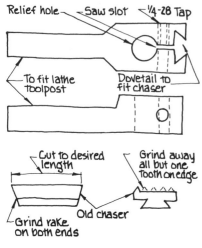

Worn or broken chasers, used in threading heads such as Landis, can be modified to make very useful single-point threading tools.

All teeth are ground off the chaser except the last one adjacent to one edge, and both ends of the chaser are ground to the desired rake angle (so that the single-point threading tool is reversible and can be used to cut up to a shoulder from either the left or the right). Use 18- or 20-thread chasers for smaller threads, and 12- or 16-thread chasers for larger sizes.

One advantage of the recycled tools is that thread form is always correct. Another is that adjustment of tool height to center is very easy because of the dovetail holder.

W E TRITZ, *Waukesha, Wis*

6.16 Recycling broken taps

Broken taps are not of much use, but they can be recycled for hand-tapping operations. A useful trick is to concentrically grind the broken end to the diameter of its proper tap-drill as an aid in starting the tap true to the drilled hole. Obviously, the proper entrance angle and clearances will also have to be restored to the reclaimed cutting portion.

HERMAN W HAUSMANN, *Weymouth, Mass*

6.17 Remove chips from blind tapped holes

The simple wire hook (A in drawing) is familiar to all machinists as a tool to help remove chips from blind tapped holes, especially small and deep ones. It usually works, as does an air gun with a small nozzle, if the chips are short and granular.

However, we often use two-flute spiral-point taps (gun taps) for materials that are difficult to tap. But while the gun tap works more easily and is less likely to break, it has the drawback of forcing stringy chips ahead of it and mashing them into a ball that's difficult to remove from the bottom of a blind

Threading

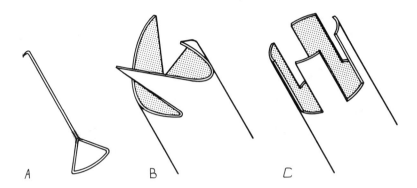

A   B   C

hole. It's true that gun taps are designed for through-hole jobs, but we like them in blind holes also.

We have, however, tried several methods for removing the ball of chips.

First, chip removal is easier if the tap is backed out after an initial pass to about half the depth and these chips are cleared before a second pass to full depth.

Drawings B and C show a couple of tube-type chip extractors that may help in difficult cases. The three-toothed tool (B) has a wall thickness of about 0.010 in. or 0.015 in. and an OD about 0.015 in. smaller than the tap-drill size (at least in the thread range from 10-32 to 1/4-20). It is simply screwed counter-clockwise into the ball of chips and then backed out—hopefully pulling the chips with it.

A more high-powered tool is shown at C. This is a piece of tubing with a few notches filed at the end. This tool is inserted in the hole, then rapped firmly with a hammer or piece of barstock. This causes the end teeth to curl inwards, surrounding the chips and allowing them to be withdrawn easily.

A little experimentation will help determine the best combinations of diameters and wall thicknesses, and a bit of filing with fine-cut needle files will generate the tooth shapes you want.

JOHN URBAS, *Cannon Tool Co, Canonsburg, Pa*

6.18 Rethreading tool for tight places

This tool is useful to thread or rethread screws in restricted spaces where an ordinary diestock just couldn't be turned. Rethreading damaged binding-post screws on electrical equipment is a specific example. The body of the tool holds a threading die. The shank is a steel tube threaded into the body. And a handle made of wood or metal is fastened to the tube. Offset pins—I used RollPins—lock the tube-to-body threads and hold the handle in place. The

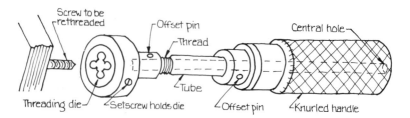

offset, and the reason for using tubing in the first place, is to allow threading of long, thin rods in the chuck of a lathe.

KEON GERROW, *Racine, Wis*

6.19 Reverse thread-cutting has its advantages

We have a number of expensive castings in which we must cut right-hand threads almost to the bottom of a blind hole. The job is done on an engine lathe that is not equipped with a thread stop, and it's a problem to prevent the tool from slamming into the bottom of the hole.

We have overcome this problem by doing the threading "backward." We use a left-hand threading tool, cutting at the rear of the bore, and feed the tool from left to right on the threading passes. This method allows the operator to position the tool right next to the bottom of the hole while the spindle is stopped. And it also provides a better opportunity for observing the tool while it is cutting.

STEPHEN EYRICH, *Hoopeston, Ill*

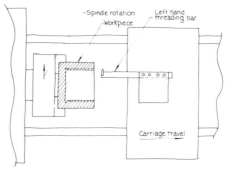

6.20 Special tap wrench eases hand tapping

Hand tapping is a very common operation in virtually any shop, but the usual T-type tap-holder, either rigid or ratchet style, leaves much to be desired. It does provide a

Threading

means for gripping the tap and a cross-bar for turning it, but it stops there. There's no provision for holding it square to the work, and it's too easy to get the tap out of alignment, and possibly break it. And the need to reverse the tap to clear chips requires release of the tool and repositioning of your hand.

The tap-handle shown ends these problems. The bottom end is made of a standard commercial tap handle with the top bored for a press-fit on the vertical extension rod. A short length of hex stock is pressed or pinned on the rod just above the tap chuck. Top of the extension rod is turned down to leave a shoulder against which the upper handle is loosely pinned or riveted so it can turn. Finally, a commercial reversible ratchet wrench is placed over the hex. The wrench should be the type that reverses by flipping a trigger—not the type that reverses upon being turned over. The wrench's fit on the hex should be loose, allowing about $\pm 1/2$ in. or more of vertical play at the end of the handle.

Use of the tool is pretty obvious, but the vertical extension really helps in holding the tap aligned, the reversing ratchet makes it easy both to turn and to reverse, and the vertical play in the handle prevents its user from inadvertently leaning down on it.

CLINT MCLAUGHLIN, *Jamaica, NY*

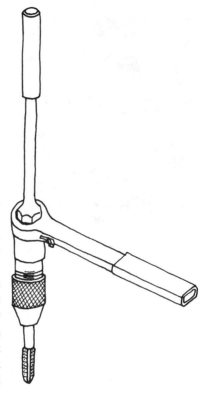

6.21 Tap-guide is spring-loaded

Standard tap-wrenches have a 60° center hole at the back end so they can be held true, for instance, by the tailstock center in hand-tapping operations in a lathe. But lack of end-float can either make it difficult to start the tap or make it very easy to break if

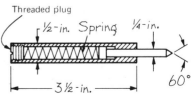

the tap has advanced a few turns into the work and the tailstock hasn't been snugged up. The spring-loaded guide shown in the drawing avoids this problem by means of an internal spring that allows float while at the same time keeping its point in the tap-wrench center hole. The photo shows it in use on an odd-angled job done recently in a milling machine.

MARTIN BERMAN, *Brooklyn, NY*

6.22 Threads milled on lathe

Large external threads on long shafts can be easily cut on a lathe by simply converting it for thread milling, which saves considerable time.

Mounted on the lathe carriage is a sturdy swivel bracket, which, in turn, holds a drive motor powering a milling cutter ground to the proper form. The swivel, of course, permits the milling unit to be set at the proper helix angle for the thread. The shaft is held between centers as usual; lead of the thread is provided by the lathe in the normal way; and the depth of cut is set with the cross-slide.

C R NANDA, *Bombay, India*

6.23 Tooling setup for multi-start threading

We have developed a tooling setup for multi-start threading that is simple, fast, and certain. And it works for double-start, triple-start, or whatever number may be required.

A rectangular steel block is bored to accept a boring bar. The bore is slotted and the block is drilled and tapped for two clamp screws, as shown. A vertical hole through the block accepts a bolt that goes into a T-slot nut to clamp the tool onto the lathe's compound. A sliding collar is made that can be clamped anywhere on the bar, and a hole is drilled in the collar that slips over a dowel pin pressed into the block, as shown. One or more U-shaped shims are also required.

Threading

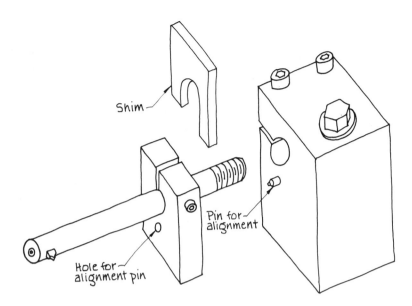

To use the tool, it is set up on the lathe in the usual manner. The sliding collar is slipped over the projecting dowel pin and clamped to the bar with its locking screw. A hex nut on the back end of the bar is used to pull the bar into desired position, and the clamp screws in the block are tightened securely. The first thread is now cut in the usual manner.

Once the first thread is complete, the two clamp screws and the hex nut are loosened and the bar is moved to the left. Now the U-shaped shim is placed between the collar and the block, the bar is again drawn against the block with the hex nut, and the clamp screws are again tightened. Now the second thread can be cut.

Key to the whole thing, of course, is the thickness of the shim, which is the reciprocal of the number of starts times the threads per inch. For example, with 6 threads per inch and 3 starts, the shim thickness should be $1/(6 \times 3)$, or $1/18$, or 0.0555 in. The first thread is cut without any shim, the second with one shim, and the third with two shims.

CLINT MCLAUGHLIN, Jamaica, NY

6.24 Trepanning broken taps

Our shop does a lot of machining of aluminum castings on small, vertical-spindle NC machining centers. The castings have upwards of 50 holes in them, most of which

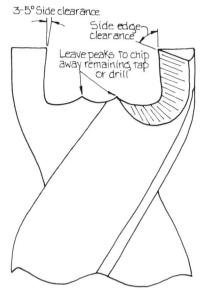

are tapped—and because sizes are small, we break a lot of drills and taps. Tap extractors are useless because we use form taps, so we initially sent out castings with broken drills or taps to have them "burned out" on EDM machines so they could be plugged and remachined. Unfortunately, however, even the smallest pieces remaining after the burning process would cause further breakage—compounding the costs.

So I designed a tool that would actually bore, or trepan, around the broken tool and chip away at the offending tip of the broken drill or tap. Using this "shell drill," I center over the broken drill or tap and then proceed to drill around it at about 200–300 rpm with straight Trim-Sol cutting oil. The next step is to try to extract the broken tool with needle-nose pliers, which is facilitated by the increased accessibility. If this doesn't work, I continue to chip away with the shell drill.

After removal of the broken tool, the hole is plugged and the part is redrilled and tapped without fear of encountering old bits of previous drills or taps. The process has saved us a great deal of time and money, and it will even work on a 2-56 tap broken off below the surface.

I make these tools by hand—using a pedestal grinder and relieving the tool on a surface grinder with a cutoff wheel. [AM received a 3/16-in. twist drill reground as shown by the author of this idea.—Editor]

ALEXANDER KASS, *Jamesville, NJ*

6.25 Tumble block for accurate thread-tools

If a threading tool isn't ground properly, it raises havoc with the threads to be cut. The illustrated tumble block, or jig, will ensure both that the 60° included

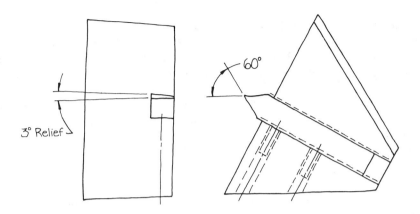

angle at the point is perfectly square with the shank and that a 3° side relief angle will automatically be produced whenever a threading bit is ground on a surface grinder.

The slot accepts 1/4-in., 5/16-in., or 1/2-in. square high-speed-steel bit blanks, which are held in place by setscrews in the two tapped holes. With the blank clamped, the holder is placed on the magnetic chuck of a surface grinder for grinding the first side, and then the block is simply tumbled to grind the other side.

Ernest J Goulet, Middletown, Conn

6.26 Use calculator for multi-start threads

Of the various methods used to cut multi-start threads in a lathe, most involve indexing the work after cutting each thread. But we do it with a pocket calculator, as follows for, say, a 12-thread-per-inch, triple-start threading job:

With the compound set at the usual threading angle of 29°, the toolbit is set just touching the workpiece, and the dials on the compound and cross-feed are both set to zero. The first thread is now cut in the usual manner. This complete, we now back up the cross-feed a few turns and re-zero the compound.

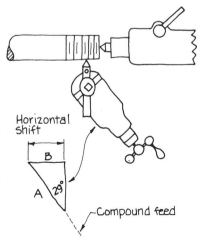

Referring to the sketch, feed motion of the compound along line A moves the tool axially along line B. What's desired is a B move equal to one pitch—1/12 in. or 0.08333 in. Since B divided by A is the sine of 29°, $A = B/\sin 29°$, the compound feed A required to move the tool to the next thread is 0.08333/0.48481, or 0.1719 in.

If the pocket calculator has trig functions, this takes less time than reading the above paragraph; if it's a simple four-function calculator without trig functions, it won't take much more time to punch in 0.48481, the sine of 29° from a trig table.

Now, by feeding the compound 0.1719 in., the tool is shifted to the next thread, and the cross-slide is fed forward until the toolbit touches the work. Write down both readings. And cut the second thread from this starting point. Then repeat the procedure for the third thread.

The reason for recording the settings instead of re-zeroing the dials is that this procedure facilitates going back to the other threads for a final cleanup pass, if necessary.

CLINT MCLAUGHLIN, Jamaica, NY

7 Tools and Tooling

7.01 Bench fixture aids accurate insert setting

For precision setting of the carbide inserts in face mills, which will ensure minimum axial runout and improved workpiece surface finishes, select a pair of Timken tapered roller bearings of different diameters that will fit the taper shank of your tooling as shown in the drawing, and mount them in the bored-out ID of a suitable length of heavy-wall tubing. Setting the outer races with Devco's Plastic Steel will permit slight adjustments for perfect alignment.

A steel arm or platform is welded near the top edge of the fixture to support a dial indicator on a magnetic base. This platform must be rigid, of course, and should extend far enough to accommodate the largest face mill in your shop.

Use of the device is obvious.

CHARLES GRAHAM, *Atlanta, Ga*

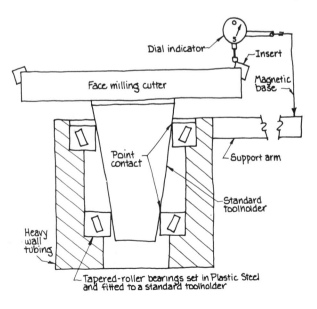

7.02 Clamp is simple, powerful

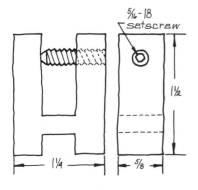

The small clamp illustrated in the sketch first found use on our portable magnetic-base electric drillpress, where it was tightened in position on a gear rack below the head as a depth stop.

The sizes are approximate and of little importance except that the web must be narrow enough to flex under pressure applied by the setscrew. The gripping surfaces were machined to a close fit on the rack.

JOHN URBAS, *Cannon Tool Co, Canonsburg, Pa*

7.03 Clamp self-adjusts for workpiece variations

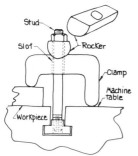

When using strap clamps it is often necessary to change the height of heel blocks to accommodate variations in the thickness of different workpieces. The clamp shown in the sketch will self-adjust for a range of thicknesses that is limited only by the length of the slot in the clamp. Both contact areas of the clamp should be rounded.

ALBERT T PIPPI, *Baltimore, Md*

7.04 Cut keyways on arbor press

The drawing shows a tool we built for cutting keyways with an arbor press. It is made of a length of steel bar with one end turned to fit the arbor press ram and the other end turned down and threaded some convenient size. A cross hole is drilled and filed square to accept a toolbit, and the end is drilled and tapped for a knurled-head screw to clamp the bit in place.

The toolbit-positioning collar is turned and bored to suit the diameter at the bottom end of the tool, and the collar is then bored off-center as shown. Half of this eccentric shoulder is finally milled off.

The toolbit is placed in its hole—note that its back end is nicely rounded—and the eccentric collar is placed over it. This is retained by an Elastic Stopnut with a lockwasher under it to give it a bit of drag. Finally, the knurled screw is inserted.

After each cut is made with the tool in the arbor press, the knurled screw is loosened and the eccentric collar is rotated slightly to feed out the bit for the next cut.

Note that the setup is for work of a specific inside diameter. When a run is to be made on holes of a different ID, it is necessary to make up a new eccentric collar of suitable dimensions.

CLINT McLAUGHLIN, *Jamaica, NY*

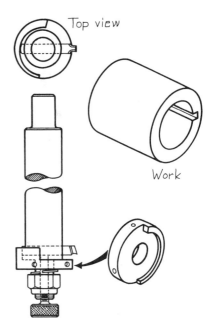

7.05 Double-coated tape fixtures plastic work

A frequent requirement in our shop is fairly precise machining of Lucite acrylic plastic. We've found that double-coated tape—sticky on both sides—can be an invaluable aid in "fixturing" these workpieces.

The first photo shows the setup for end-milling six 3/8-in.-wide slots 0.030 in. deep in 1/16-in. clear Lucite. The blank was lined up with two pins in a

T-slot after the tape was laid down on the milling-machine table, after which the blank was secured and the pins were removed.

The second job required 0.016-in.-wide slots 0.187 in. deep. Here the work was double-taped to an aluminum block clamped horizontally in a table vise. First step was to mill off the top surface of this block true and horizontal, leaving a "rail" along the rear edge for aligning the workpiece. Note the lack of clearance below the milling arbor for any conventional clamps or bolts.

The third situation required us to produce 0.187-in.-wide strips of Lucite to close tolerance from 1/16-in.-thick material 8 in. long. An 0.040-in. slitting saw was used for this in a vertical spindle. Again the fixture incorporated an intermediate surface, this time made of scrap 3/8-in. Lucite and clamped vertically in the vise, with double-coated tape to secure the workpiece.

MARTIN BERMAN, *Brooklyn, NY*

7.06 Double-duty Allen wrench

Typical of most tool and die shops, we use T-handle Allen wrenches a great deal. We make our own, however, with some special features.

We take 3-in. lengths of straight Allen-wrench hex stock and press them into round stock in which an undersize hole has been drilled axially. Hole diameter should be the dimension across flats plus half the difference between the flats dimension and the corners dimension.

A 4-in.-long T-handle is welded to the end of the wrench body. This handle is also drilled axially with a hole large enough to accept the body of the next size wrench. This allows the wrenches to be used in tandem, as shown, for increased leverage for breaking loose or tightening socket screws. And this increases the number of applications for which the T-wrenches can be used. They've proved extremely versatile.

MICHAEL P SCHMIDT, *Owatonna, Minn*

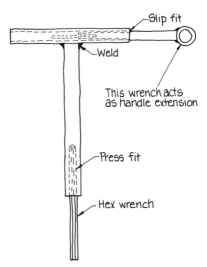

7.07 Double-jointed vise stop

Here's a simple, practical, and inexpensive work-stop for a milling machine vise. It's made of two steel blocks (measuring 3/8-in. x 1/2-in. x 1 1/2-in.) and a length of 3/16-in. rod, with the parts machined and assembled as shown.

The side of the machine vise must also be drilled and tapped 1/4-20 for mounting the work-stop, which is easily adjustable to a variety of heights and angles and is quickly removable for those jobs when it's not needed.

GENE BRIGHTHAUPT, *Newark, Del*

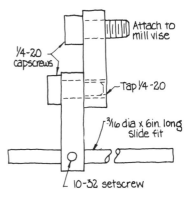

7.08 Fixture for turning, grinding eccentrics

The accompanying drawing, in somewhat simplified form, shows the basic idea behind a turning, grinding, or inspection fixture that features eccentricity continuously adjustable from zero up to whatever maximum the specific dimension of the fixture allow. The fixture itself would typically be held in a three-jaw self-centering chuck.

The unit consists of two nesting cylinders, both bored eccentric by precisely the same amount (dimension "a") and with the face calibrated through 180°.

A simple trigonometric formula, which follows, is used to convert the angular calibration into the linear dimension of the eccentricity or offset.

Eccentricity = 2a x cos $1/2\,\theta$

in which 2a is the maximum possible eccentricity and θ is the rotation angle indicated on the face calibration.

P S KULKARNI, *Hubli, India*

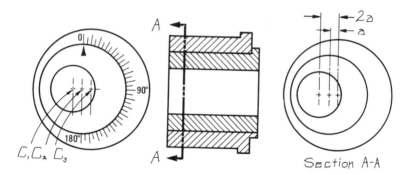

7.09 Flycut that keyway

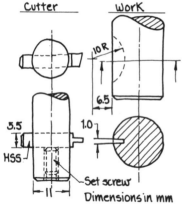

Machining some shafts in our shop required Woodruff-style keyways, which, in turn, would have required special cutters of 20 mm diameter and 1 mm thickness. These were just not available, and it would have been very costly to make them.

We solved the problem simply by inserting a small high-speed-steel toolbit in a shank, as shown in the sketch, which was used to flycut the keyways. The simple tool was made in about an hour, and it probably was eight to ten times cheaper than a saw-type cutter would have been.

C P R VITTAL, *Andhra Pradesh, India*

7.10 Guide plates for deep EDM operation

The illustrated setup was used to EDM-machine a large weldment in our plant (the workpiece is drawn in a much-simplified form). The guide plates were specially made to suit the job and were located very precisely in the workpiece with nonmetallic shims on both sides of each guide plate and at both the top

Tools and Tooling

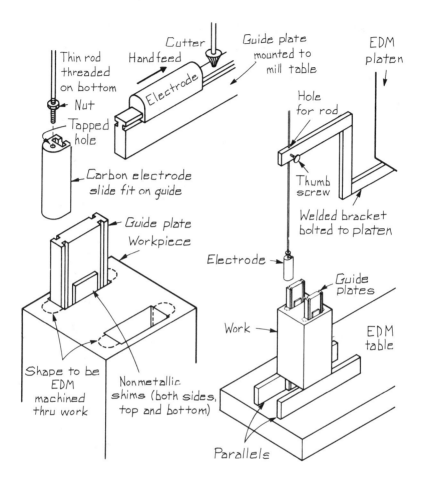

and bottom ends. The shims were nonmetallic in order to insulate the guide plates from the workpiece.

It was also possible in our case to finish-machine the carbon electrodes (an adequate number were prepared in advance) using a guide plate as a fixture. As shown in the sketch, the guide plate was mounted on a milling machine table and was used to guide the electrode as it was manually fed past the form cutter.

Both guide plates—which should be longer than the workpiece depth and which should project both above and below the work—were set in place before starting the EDM operation. Although it might be possible to feed all four electrodes (in the illustrated workpiece) at one time, we used only one at a time.

Electrode feed was accomplished by means of a 1/8-in. vertical rod screwed into a tapped hole in the end of the electrode. Upper end of this rod was affixed by a thumbscrew to a welded bracket bolted to the platen of the EDM machine. The small diameter of the rod allows it to flex somewhat during feeding to accommodate any misalignment in the setup.

When the electrode has been sunk into the work as far as head travel permits, the feed-rod thumbscrew is loosened, the EDM head is raised, and the thumbscrew is retightened. This sequence can be repeated several times to permit EDMing through very long workpieces that exceed the head travel limit of the EDM machine.

Electrodes are replaced as necessary, and changing them is quite simple.

JOHN URBAS, *Cannon Tool Co, Canonsburg, Pa*

7.11 Hobbing a spline on an overlong shaft

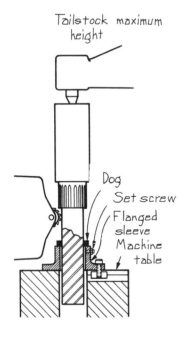

We were confronted with the problem of hobbing a spline on a shaft that was several inches longer than the capacity of our hobbing machine. Our solution was to lower one end of the shaft through the hole in the center of the turntable.

A flanged sleeve was made with a bore large enough to allow the workpiece through, and four equally-spaced radial holes were drilled and tapped 1/2-13 for setscrews and spotfaced for jam nuts. Three holes were drilled through the base flange to line up with the T-slots in the machine table.

With the tailstock raised to the limit, the shaft couldn't be loaded with the sleeve clamped to it. So we added a grinder dog above the sleeve to prevent the shaft from slipping through, and we used the setscrews in the sleeve as a four-jaw chuck to align the shaft and hold it for hobbing.

With a short feed per revolution and shallow cuts, the spline was hobbed to meet specifications.

DUNCAN VERTREES, *Ypsilanti, Mich*

7.12 Miniature setup jack

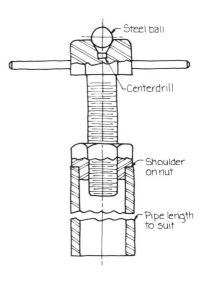

A miniature machine jack is indispensable in many setups, and this one is both easy to make and highly adaptable. The parts are a hex-head capscrew with the head center-drilled for a steel ball and optionally cross-drilled (as shown) for a torque handle, a matching nut with a turned shoulder, and a length of pipe suitable for the job at hand.

The steel ball pivots in the drilled hole and slightly indents the workpiece, which prevents the jack from moving under load. The turned shoulder on the nut pilots itself into the end of the pipe, both ends of which should also be turned square, and provides a well-defined bearing surface.

A bit of grease or oil on the ball and on the threads makes the jack easier to turn. And to adapt the jack for other jobs, all you need is a new length of pipe. You'll soon find you have a collection of them.

KEON GERROW, *Racine, Wis*

7.13 New life for old C-clamps

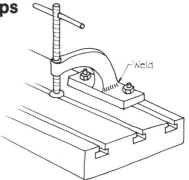

The drawing shows a simple, but very handy clamp that we usually use in pairs for various machine setups. Ours are made of damaged C-clamps, which are cut off as shown—using an abrasive cutoff wheel, if necessary—and then welded to a base made of barstock or plate with the mounting holes drilled in the base before welding.

JOHN URBAS, *Cannon Tool, Canonsburg, Pa*

7.14 Non-tipping bench blocks for spacing

A pair of bench blocks can be very handy for driving pins out of assemblies, for resting dies on, for through-drilling under a drillpress, and for literally dozens of other similar tasks. But, when they're set up on edge, at least one of them always seems to get knocked over.

Not this simple set, however. I use a pair of 1-in. x 2-in. x 3-in. blocks, although any size could be made. The drawing is completely self-explanatory.

RON STANWICK, *Englishtown, NJ*

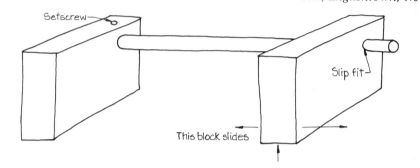

7.15 Roller tool straightens bent shaft in lathe

To straighten some bent shafts, we made the roller tool shown in the sketch, which incorporates a pair of shielded ball bearings as the rollers.

With the tool clamped in the toolpost and with a bent shaft mounted in the lathe, the machine is started and the shaft is heated at the bend with a torch. When the shaft is red hot, the cross-slide is fed in until the rollers contact the work—not at the red-hot area but near it. The reason for avoiding contact at this hottest point is that rolling it there will reduce the shaft diameter.

By cross-feeding the roller tool against the rotating shaft, it's relatively easy to perform the straightening operation. But one caution must be observed: Heating the shaft causes it to expand; thus the tailstock must be carefully backed off during the operation to limit the axial load against it.

LEE C WILKERSON, *Keyesport, Ill*

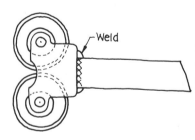

7.16 Screw locates and clamps

It is customary in designing or building fixtures to position parts over close-tolerance locator pins and then clamp them in a separate operation. Many a problem, however, can be solved by using a hole chamfer in a part as the surface for both locating and clamping. The drawing shows a simple drilling fixture as an example of the technique, but it can be applied to milling fixtures, gaging setups, and other mechanisms as well—using one or more screws as needed.

WILLIAM SLAMER, *Menomonee Falls, Wis*

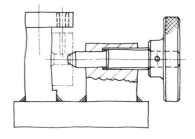

7.17 Standard bushing is a dowel in tight places

A hollow "dowel" can often save the day when there's just no place to put a locking screw; you just put the screw through the dowel. And because the ID is of no great significance as long as it's large enough for the screw, you can use a standard, headless, press-fit bushing for the job, reaming the mating pieces to fit the bushing's OD, of course.

ERNEST J GOULET, *Middletown, Conn*

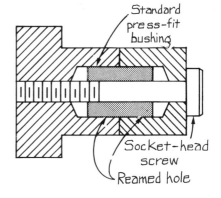

7.18 Standard racks make multipart fixture

The accompanying drawing illustrates a multipart fixture that has proved very useful in our grinding and milling department. The clamping jaws are made of standard, commercially available steel or brass racks, which are made into matching sets for mounting on the sliding holder. Several pitch sizes of racks can be made interchangeable so that an even greater variety of rod sizes can be accommodated, although even a single set will accept a range of diameters. Overall length of rack sections and number of teeth will, of course, depend primarily on whether long-run or short-run production is envisoned.

Clamping action of the rack teeth is very firm and location is positive because the fixed rack's tapered tooth form acts as a narrow-angle V-block while the sliding rack comes in to contact each part at a third point.

FRED STAUDENMAIER, *Bedford, Quebec, Canada*

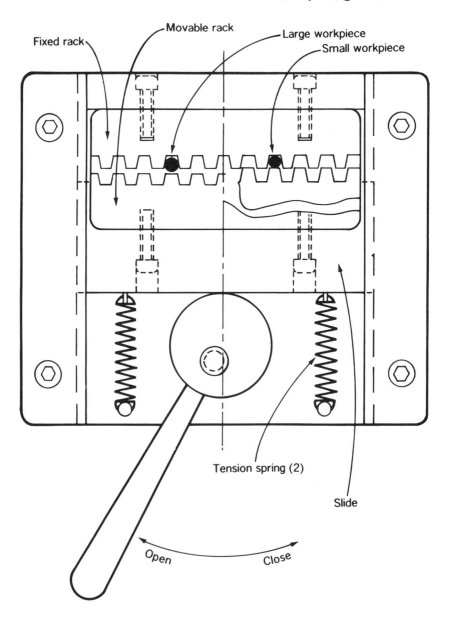

7.19 Stop 'saves' position

Whenever a long, round workpiece requires operations on both ends that must be aligned—such as a long, off-center through-hole that has to be drilled from both ends—attach a stop to the work to preserve radial alignment.

For a short run of parts, the stop might typically be leveled with gage blocks or an indicator. But we simply rotated the stop until one corner rested on the machine table. After work on the first end is complete, the workpiece with the stop still clamped in place is turned end for end, and the stop is again rotated into contact with the table—making sure that the same corner is used.

JOHN URBAS, *Cannon Tool Co, Canonsburg, Pa*

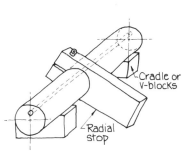

7.20 Strap hinge 'handle' for grinding screws

Machine screws can be shortened by hand grinding on a pedestal or bench grinder with the aid of this simple holding "fixture." Merely drill and tap a pattern of various commonly used machine-screw sizes in both halves of a strap hinge, as shown in the sketch, to accommodate the screws.

The screws can be force-ground without the hazard of getting burned fingers as when hand-holding the screws. Moreover, it will not be necessary to file off the rolled-over end burr and thread because it is straightened as the screw is unthreaded from the hinge.

ALBERT T PIPPI, *Baltimore, Md*

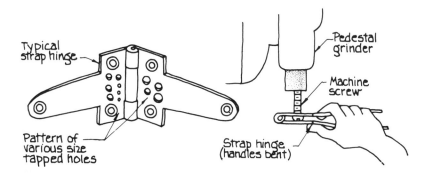

7.21 Tips for 'adjusting' holes

You can make an end mill cut a larger hole or slot than its normal cutting diameter by slipping a piece of paper against one side of the shank before you slip it into the holder or collet.

You can make a reamer ream a larger hole by scraping each one of the flutes with a piece of carbide against the cutting edge. The harder you scrape the edges, the more you will increase the size of the hole.

And if a hole to be drilled and reamed was drilled a few thousandths off location, you can set up the workpiece at an angle—up to about 30°—and countersink it on the side of the hole in the direction you want to move the hole. Setting it up flat again, a reamer will pull towards the countersunk side of the hole.

GENE BRIGHTHAUPT, *Newark, Del*

7.22 Two uses for standard angles, channels

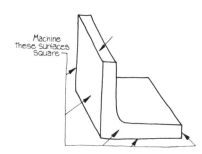

Angle iron is a hot-rolled product that doesn't need any stress-relieving to be virtually non-deforming. A length of this material, which is available in many standard sizes, can be made into a handy angle plate for such jobs as surface grinding, or it can be machined as needed for use as a fixture. Just be sure to machine it square.

And short lengths of standard channel can quickly be converted into stepped strap clamps by drilling a hole for the T-slot bolt and cutting off one leg to suit whatever workpiece you want to clamp onto a milling-machine table.

BEN SCHNEIDER, *West Orange, NJ*

7.23 Use strap clamps for parallel clamps

When machining two or more parts together, there never seem to be enough suitable clamps at hand. C-clamps are often awkward for machining or setup, and not much else is available.

The sketch shows how a pair of strap clamps can be used to make a parallel clamp. These will provide greater clamping force than C-clamps, and have the added attraction of a low profile as an added plus.

JOSEPH D JUHASZ, *Michigan City, Ind*

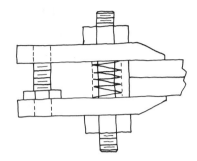

7.24 Vise jaws extended

It happens sometimes that a round or irregularly shaped workpiece must be held in a particular vise the jaw-height of which is less than the radius of the work. The self-explanatory sketch illustrates how the vise jaws can be extended by using three parallels, one of which must be exactly equal in width to the diameter of the part being clamped. This parallel serves the same purpose as a heel block or shim block under the outboard end of a strap-clamp used on a T-slotted table.

FEDERICO STRASSER, *Santiago, Chile*

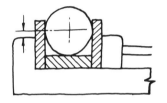

7.25 Work clamped with eccentric capscrews

Very simple and effective fixtures for small workpieces can be made by clamping with socket-head capscrews with their heads ground off-center. As the screws are rotated into the work, the eccentric heads force the work against the locator pins and simultaneously pull the work down against the fixture plate.

These screws can be made up in batches and stocked for use when needed. The amount of eccentricity depends on the job at hand; a low degree of eccentricity produces greater clamping force, but allows for smaller variation in the size of the workpieces.

CLINT MCLAUGHLIN, *Jamaica, NY*

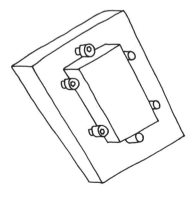

8 Forming and Press Tooling

8.01 Access holes make it easy

Don't forget to drill an access hole in line with that hard-to-reach cap screw, such as when you're fixing a chute against the bolster of a multislide machine. You can eventually tighten the screw with quarter turns of your Allen wrench. But it's a lot easier to slip the wrench through the clearance hole and twirl the screw up tight in a jiffy.

BEN SCHNEIDER, *West Orange, NJ*

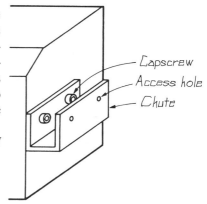

8.02 Accordion die telescopes to save material

When you put different sized parts in a progressive die with interchangeable inserts, there has to be a waste of material. The die must be designed to suit the largest part, so the material required to produce any smaller parts must leave more waste in the strip skeleton. The reason, of course, is that the die lead or progression is usually a fixed dimension.

We had to make some quantities of 1/16-in.-thick dished copper washers with various outside diameters, but with the same form and center hole ID. The cost of copper prompted consideration of separate dies for each, but production forecasts weren't high enough.

Instead we designed a die with interchangeable spacers between stations so that lead could be varied to suit the individual part. The drawing shows the bottom half of the die with a clamp to sandwich in the die sections. The punch holder (not shown) is similar. The stripper is also interchangeable to allow

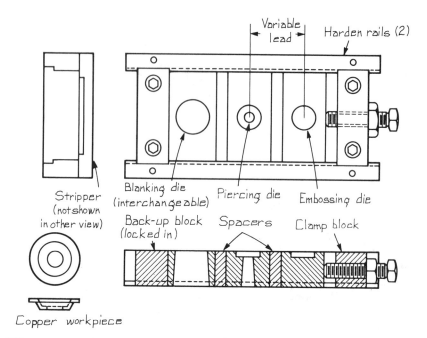

different stock widths, and the holes in the die shoe must be either elongated or large enough for clearance.

ERNEST J GOULET, *Middletown, Conn*

8.03 Add 'feet' to stabilize swaging dies

We manufacture various diameters of swaged chimney pipe. To prevent difficulties resulting from the smaller-diameter inner swaging dies from tipping

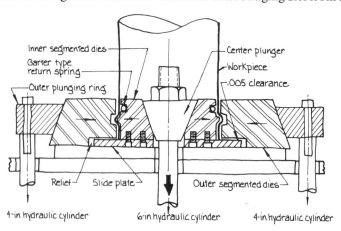

during the operation, we added a simple slide plate to enlarge the base of each one of these inner die segments, as illustrated. A relief was machined in each outer segment to provide space for the plates. As well as preventing the possibility of the dies tipping as they expand outward, the slide plates also act as locating surfaces.

MICHAEL SCHMIDT, Detroit Lakes, Minn

8.04 Adjustable die forms channels from sheet

Our model shop frequently has to make channels from sheet metal in odd and narrow sizes, some of which are impossible to form on a press brake with conventional tools. To fill this need, I devised a simple tool that can be adjusted easily to produce a variety of sizes. And it produces channel shapes to reasonably close tolerances.

The tool consists of two heavy steel bars doweled together with a press fit in one and a slip fit in the other. One bar is drilled and tapped in two places and the other has matching clearance holes for a pair of bolts to hold the bars together at proper spacing. The top edges of the bars are given a fair radius. One bar is fitted with an adjustable stop to ensure that the blank's edge will be parallel with the die.

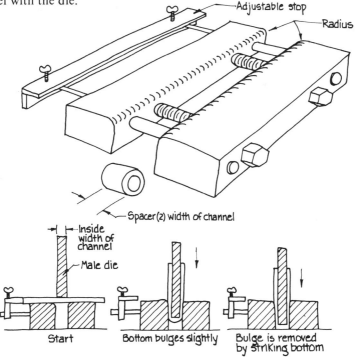

To use the tool, it is necessary to determine the outside dimension of the channel. If more than one channel is to be made, it is advisable to make spacer bushings to that length and slip them over the bolts, which are then tightened on them to establish the size of the channel. If only one channel is to be made, it's usually sufficient to simply tighten the bolts until the correct spacing is achieved.

A male die is made to the inside dimension of the channel, and its bottom edges are rounded slightly. I coat the bars with a thin layer of lubricating grease and lay the sheet metal over them. After I have centered the male die, I press it into the work with a hydraulic press or our press brake. The die should bottom, which will flatten the channel. The formed part may be removed by loosening the bolts or by pressing the male die right on through. If there are many pieces to be made, it's advisable to use the press brake to hold the male die.

The size of the bars that have proven most useful in our shop are 1.00 x 2.00 x 18 in. laid flat for maximum resistance to deflection. The dowels and bolts are 1/2 in. diameter.

ART DRUMMOND, *Walworth, NY*

8.05   Bevel extends punch life

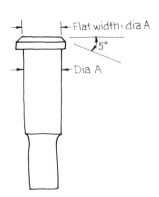

Sometimes when operating through the rear of a Multi-Slide die the head of an auxiliary punch will snap off. And this can create a jam in the die and possibly other damage, as well as necessitating replacement of the broken punch.

Such breakage is often due to improper guidance of the auxiliary punch through the die and an out-of-square force against the head of the punch by the auxiliary slide. The off-center force can be relieved by bringing the center of pressure into line with the body of the auxiliary punch, and this can be done simply by beveling the punch head as illustrated.

BEN SCHNEIDER, *West Orange, NJ*

8.06   Bolt as small jack opens jammed dieset

We recently had a forming dieset jam together because of some stock that was oversize in thickness. With only 4 1/2 in. between the top and bottom plates, available hydraulic jacks could not be used.

A 1-3/4-in.-dia bolt and nut was inserted between the two plates—with a thrust washer between the nut and the die plate to reduce friction—and the nut was turned off the bolt to force open the jam. To reduce cocking, two or more bolts could be used.

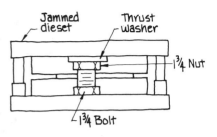

GERALD L CLAY, *Alliance, Ohio*

8.07 Broken-punch detector stops operation

Using horizontally mounted hydraulic cylinders, we were piercing six holes in a sheet metal cover. One punch, smaller than the others, would occasionally break even though it was well guided.

An even worse problem, however, was that the operator did not always notice when the punch broke, and we would accumulate a load of covers that had to be worked for salvage.

Because of the fragility of the small punch we had to live with the occasional breakage, but we didn't have to manufacture bad parts. We solved the latter problem by devising a broken-punch detector that would automatically shut down the machine.

The actuating cylinder was moved back for space, and a yoke was added between it and the punch assembly. A crossbar was fixed to the piston rod and another was made to slide over a threaded punch-holder. Two curved bands, or straps, of spring-tempered steel, were used to link these crossbars, as illustrated.

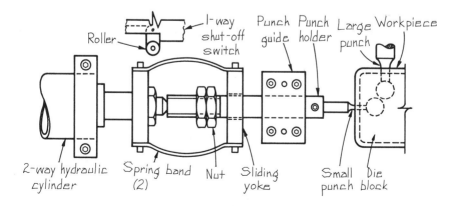

Flexibility of the bands allows them to straighten out under the tensile load imposed by the stripping action. But when the punch is broken, there's no stripping load, and the spring bands retain enough of their bow to actuate the shutdown switch.

ERNEST J GOULET, *Middletown, Conn*

8.08 Built-in 'jacks' simplify disassembly job

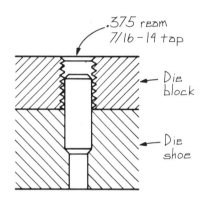

Handling heavy dies for maintenance was a problem in our shop. Diesets were too large to turn over easily, and components were awkward to remove to lighten the job.

Our first step was to bolt all blocks from the inside of the die, which helped some, but dowels still left some problem areas.

By cutting 7/16-14 threads in 0.375-in. dowel pin holes as shown in the drawing, we were able to put jackscrews directly atop the pins. Because the dowel holes were not reamed all the way through the dieset, the pins do not push out the bottom, but rather allow the block to be pulled off the pins.

Checking standard tap-drill charts shows that there are other combinations that will work, but this particular combination seemed to be most satisfactory in our case. We still maintained 37% of the original bearing surface as compared to a standard hole, and the thread depth was 67%.

KENNETH ELSHOF, *Grand Island, Nebr*

8.09 Cammed stock guide prevents jamming

Punching heavy stock often means trouble, especially if the part or the strip is relatively narrow. This may be caused by the natural tendency of the stock to spread under the punching action—sometime by several thousandths of an inch. This increase in width added to the commercial tolerance on the width of the stock may easily exceed the allowance in the width of the nest. If the nest is too narrow, the stock may jam; if the nest is widened to accommodate, location tolerances will suffer.

The solution consists of making the nest self-adjusting, as sketched. One side is fixed, the other adjusts. A couple of light springs (not illustrated) keep the movable slide open to its maximum width position, so the stock (or blank) can be fed easily. When the press is tripped, the ram descends and the cam drives the roller-equipped lever against the slide to firmly position the stock against the fixed edge. As the ram continues downward, the punch does its job.

Any expansion of the workpiece caused by the punching simply pushes the slide, lever, and cam against the compression spring. And as the ram rises, the pressure is released so that the finished stamping can be easily removed from the die.

FEDERICO STRASSER, *Santiago, Chile*

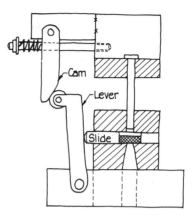

8.10 Combination die bends, then cuts wire part

A new product required a part to be formed of 0.090-in. copper wire to tighter than normal tolerances of ±0.002 in., which could not be held with a forming die of conventional design because of the behavior of the material. A simple combination die that formed and cut the wire in a single operation did the trick.

As shown in the drawing, the die has a slot and two end stops, which hold the wire (precut 13/64 in. longer than overall finished length). As the press ram descends, the forming punch shapes the "U" first, and as the ram continues downward the two cutting punches shear the wire accurately.

The cutting punches are rather large in diameter to minimize the concavity at the ends of the wire, which is hardly noticeable

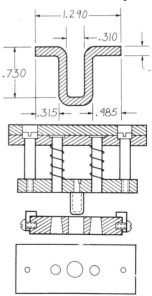

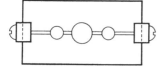

anyway because of the relatively small diameter of the wire.

The die is also provided with an ejector (not shown), and the entire tool was mounted in a conventional dieset. It did the job, however, meeting the tight tolerances easily, and doing it at a highly satisfactory rate of production.

FEDERICO STRASSER, *Santiago, Chile*

8.11 Combo tool embosses, punches, blanks

The sketch shows a combination tool for producing washer-like components in which a circular reinforcing bead (or rib) is embossed between the inner and outer peripheries. With the sheet stock in place on the lower, fixed die, the ram descends to perform successive operations by means of the telescoping action of the upper tool members.

First, the reinforcing bead is embossed by the combined action of the stationary die and the spring-loaded intermediate sliding member.

While the stock is firmly held between the two embossing dies and the compression spring yields to further descent of the ram, the central punch produces the hole in the workpiece.

At the same time, the outer ring—in combination with the outer periphery of the stationary die—trims or blanks the outer contour of the workpiece.

As the press ram continues downward a little farther, two chisel-like supplementary steel bars mounted on the die-holder base sever the strip skeleton at its narrowest point against the bottom of the outer movable ring. The scrap is thus split into two sections, eliminating the need for a stripper.

For this particular workpiece, a combination tool was essential because the

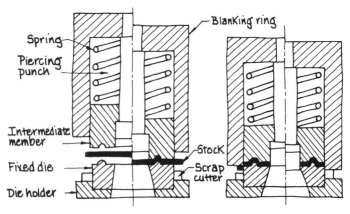

reinforcing bead has to be embossed prior to punching and blanking, which is made necessary by the proximity of the bead to both the internal and external edges of the part. Distortion would have resulted if the embossing were done in a separate tool afterwardss

FEDERICO STRASSER, *Santiago, Chile*

8.12 Custom components with standard tooling

To produce a one-piece order for a tapered-box housing of a nuclear handling mechanism, we used an air-bending die, as shown in the sketch, instead of a customized punch and die. The approach saved the cost of what would necessarily have been some very expensive tooling, and a further economy was realized by our use of a long-dormant large-radius lower die that we already possessed. The only tooling purchase necessary was the punch with its gooseneck clearance.

Made out of 1/2-in.-thick 304 stainless steel, two identical tapered "channels" were first formed as indicated. Excess material was then machined off of these, and they were welded together to produce the desired component.

GEDEON O TRIAS, *Rockford, Ill*

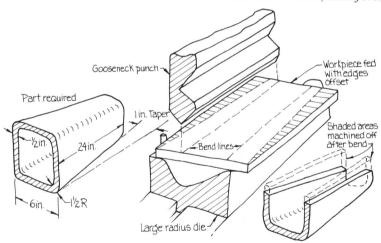

8.13 Deflector solves punch-breakage problem

A U-shaped stamping called for concentric holes in each of the two legs to suit a close-fitting pivot pin, as shown in simplified form at the left of the drawing. To meet the concentricity requirement, the holes were simultaneously punched with the tooling setup indicated after the workpiece was formed.

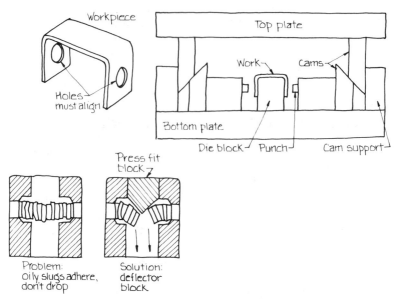

Problem: Oily slugs adhere, don't drop

Solution: deflector block

A difficulty arose in the central die block, however. What happened was that the relatively small and lightweight slugs, coated with oil, would stick together and not fall free. With this building progressing from both sides of the die block, the slugs would eventually leave inadequate clearance between, preventing the next punching operation and breaking the punches. A counterbored clearance in the rear of each side of the die block, incorporated to facilitate easy movement of the slugs, did not prevent the problem from arising occasionally.

Ultimate solution of the difficulty proved quite simple: a deflector block with a V-shaped bottom was pressed into the opening at the top of the die block, and this effectively breaks the oil adhesion to allow the slugs to drop free before the buildup becomes critical.

RAJN S KENY, *Bombay, India*

8.14 Die-holder accepts workstops, guides

An equally spaced grid of tapped (10-32) holes in the die-holder of our DiAcro sheetmetal utility punch has added significant versatility to its use. Using these tapped holes to attach workstops and positioning guides greatly simplifies and speeds use of the equipment for short-run production with good tolerances.

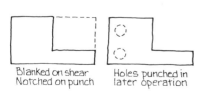

Blanked on shear Notched on punch

Holes punched in later operation

For example, the setup shown was used to produce the part sketched. First, rectangular blanks were cut on a shear. Then the large notch was punched on the DiAcro in the setup in the photograph. Finally, after a sufficient number of the L-shaped parts were notched, the DiAcro setup was changed to allow punching the two small holes. Workpiece positioning was a snap for both operations.

MARTIN BERMAN, *Brooklyn, NY*

8.15 Die inserts cut costs in press brake tooling

Low-volume parts bent between 90° and 175° on a press brake require female dies of the proper width for the required bend. Instead of buying a universal adjustable die for these sizes, we made a box section from low-carbon steel that accommodates removable pads of 1045 bar stock that are surface-hardened. For different parts, new inserts can be made of the proper thickness, or, even simpler, the existing inserts can be shimmed to the proper spacing.

GARY R GREGG, *Shippensburg, Pa*

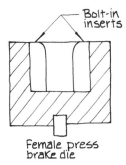

Female press brake die

8.16 Die offset is adjustable

Producing offset configurations in sheetmetal is common practice in all fabrication job shops, and standard offset dies are available in increments of 1/8 in. ID to OD. But parts are often designed without regard to material thickness or standard dies, which typically requires either expensive special dies or a double operation in which the part is flipped over and very narrow 90° dies are used.

The use of shims, as shown, has enabled me to satisfy all dimensional requirements

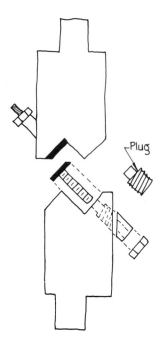

with standard dies already in the shop. For prototypes or short runs, I just use double-coated tape or an adhesive to hold the shims in place during the forming operation. For production runs, I have drilled three holes in each offset die, and the shims have matching 8-32 weld studs (or press studs) that protrude through the offset die to accept angled washers and securing nuts. The lower sketch shows this "exploded," the upper sketch shows it assembled.

The thickness of the shim or shims will generate the required offset. By using standard material gages, or combinations of gages, a range of dimensions can be set up. The shims are marked and stored for future use.

The last 1/4 in. of the through holes is tapped 1/4-20 for installation of plugs (as shown) to prevent marking of workpieces formed by the dies when they are used in their original configuration without shims.

DUANE C HARDING, *La Habra, Calif*

8.17 Die produces three cups per stroke

The principal objective in the design of this double-action blanking-and-cupping die—which produces three cups per stroke of a single-action press—was to maintain positive control during the blanking operation. Key feature of the design, which works very well, is the guided and spring-loaded plate that acts as (1) the blanking-punch holder, (2) a pressure-pad or stripper plate, and (3) a guide-plate for the draw punches.

Here's how it works: The guided plate is positively and firmly locked in the cutting or blanking position by the two lever arms. After blanking, however, and during the continued downward travel of the ram, the lock arms are cammed outward by their actuators. This action releases the positive driving action of the lever arms so that the plate is now driven by the springs—it's now a spring-loaded pressure pad, in other words, and will now push the blanked disks (by means of the draw punches) through the blanking dies and onto the drawing rings, finally applying the spring-loaded blank-holding pressure for the draw. And now the draw punches complete the job by pushing the blanks through the draw-die rings.

Forming and Press Tooling

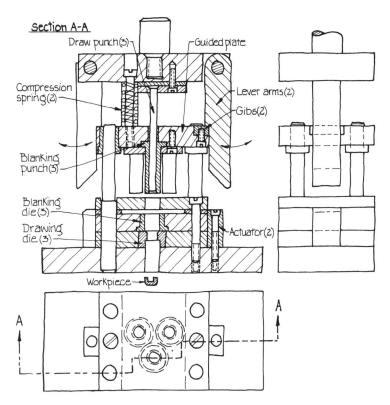

As the press moves back up again, the springs expand and move the guided plate into starting position (controlled by the stripper bolts), and gravity moves the lever into their lock position.

The die runs nonstop with an automatic strip feeder attached.

FRED STAUDENMAIER, *Bedford, Quebec, Canada*

8.18 Double-acting hand-press tests methods

For the prototype department or experimental shop of a company making small stamped components—primarily eyelet-type workpieces—a small, special, shopmade manual double-action press has given excellent results for trying out the feasibility of producing small components that require some blank-holding action in production (drawing, riveting, assembly, etc.). The accompanying sketch shows the press diagrammatically. It has two telescoping spindles; the outer, larger one actuated by a spiral or eccentric cam, and the inner one by a toggle-link arrangement.

There is a sturdy body on which are mounted three bushings and the

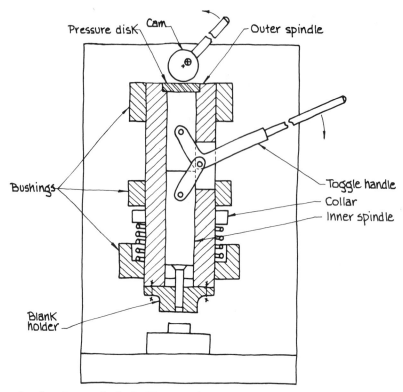

bearing for the cam's shaft. The bushings provide the "housing" for the outer spindle, on which a ring-shaped collar is provided. The ring serves a double function: it acts as a stop for the upward stroke, and it provides a bearing surface for a compression spring that powers the upward return motion.

Inside of the large spindle is a bore for the small spindle, which is moved positively up and down by means of the toggle-link mechanism, the levers of which pass through a slot in the side of the outer spindle. The fulcrum of the actuating lever is in the outer spindle itself, so that the fulcrum rides up and down with the outer spindle.

In operation, the outer spindle carries the blankholder. Its pressure adjustment is accomplished either by changing the shims of the stationary die to alter shutheight, by rotating the spiral cam to a new position on its shaft, or by repositioning the cam's height by means of an adjusting screw. The working punch is actuated by the inner spindle, to which it is joined by means of a suitable coupling screw or nut.

The dies in question are extremely simple tools, of course, with only the bare essentials in accordance with sound short-run practices.

FEDERICO STRASSER, *Santiago, Chile*

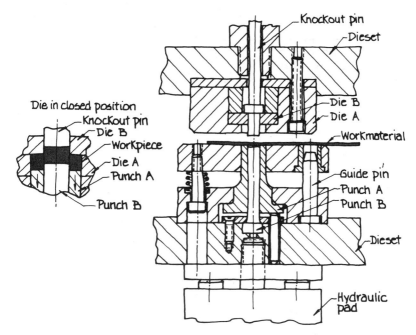

8.19 Double-action in a single-action press

Many metal stampings that call for special considerations of straightness, cutting direction, or concentricity of inner and outer periphery are normally made either in an expensive progressive die or in a double-action press.

An alternate way to perform such operations—and we have used this method for some time—is to use a single-action press to which a hydraulic pad has been added to provide sufficient resistance for the force necessary to punch the outer periphery of the work.

Operation is simple: As the press ram descends, it cuts the outside shape of the part with punch A and die A against the resistance of the hydraulic pad. As punch A completes its work and comes to the dead point, punch B is cutting against die B.

As the press again moves up, the workpiece is ejected by means of the knockout pin and an air blast.

MOSHE HARARI, *Givataim, Israel*

8.20 Dummy pad aids strip feed accuracy

For parts that are both notched and formed in a progressive die, especially if the part is made of thin stock, the forming operation sometimes pulls on the stock and causes a variation much like that which would result from an

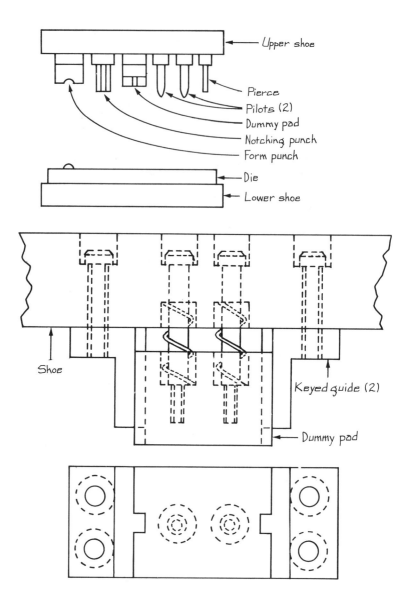

inaccurately set feed or inaccurate piloting if pilots are used.

One possible answer is to install what I call a "dummy pad," which does no more than clamp the strip in place before the forming operation takes place. The feed should, of course, have reached the end of its stroke before the dummy pad comes down to lock the strip as the die closes. Also the pad should not lock the strip before the pilots, if used, locate the strip.

It is necessary to use keyed guides, or supports, because the stripper bolts can't provide dependable guiding.

BEN SCHNEIDER, *West Orange, NJ*

8.21 Economy stop feeds narrower strip

Unlike the widely used French stop, which requires extra-width strip for punching out blanks, the stop illustrated here permits you to economize by feeding strip that's just barely wider than the blanks you're making. In most cases there is sufficient stock area between blanks for the necessary notches.

Operation of the stop is simple. The right side of the notching punch produces a dead stop for the spring-loaded pawl to catch on. At the same time, the left side of the punch cuts out an inclined ramp for the pawl to ride up and over to the next notch.

BALLARD E LONG, *Oak Ridge, Tenn*

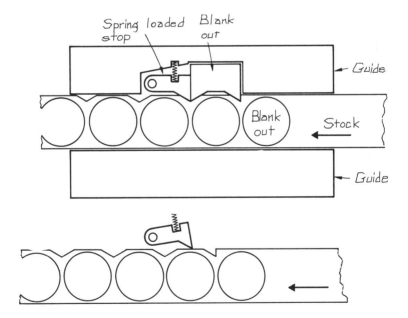

8.22 Embossing die flares holes from inside tube

Two flared (embossed) holes had to be formed in the ends of short lengths of tubing, as shown in the upper drawing, which required tooling with some unusual features.

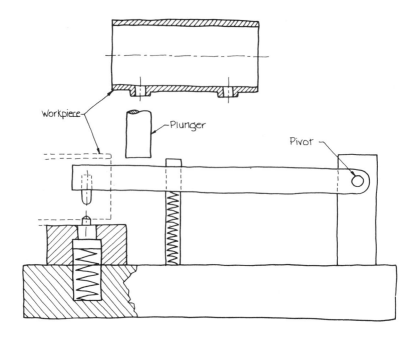

Because the tubing was relatively large in diameter, there were no problems in punching the pilot holes with a standard horn die. But the flaring job couldn't be done with a gooseneck punch because the hole was too deep and the long press-stroke would have caused extra difficulties.

So we made an indirect-acting horizontal pivoting punch, which is actuated by a plunger that depresses the punch when the press ram descends. With the ram in the rest (upper) position, the punch is lifted by a spring-loaded pin, and the upper position is limited by a simple hook stop.

The workpiece-locating stop is a spring-actuated pin that moves vertically in the bore of the die-plate. It is so shaped and dimensioned that it also acts as the ejector.

In operation, the prepunched part is introduced into the die and the hole is placed over the stop-pin. The press is tripped, and the plunger depresses the pivoted punch-holder bar causing the punch to depress the stop pin and perform the embossing operation. When the ram ascends, the punch-holder is lifted by the spring-loaded pin and the stop-pin/knockout ejects the workpiece from the die plate.

The tool has worked quite well, especially because the long punch-holder arm results in a near-straight-line motion of the punch.

FEDERICO STRASSER, *Santiago, Chile*

8.23 Ending pullup with stainless steel slugs

Stainless steel has the property of wearing die cutting edges fast. And frequent resharpenings quickly take the die down below its land, resulting in too much clearance. Then the slug pulls up, jams the strip, and may even shear the punch.

One way to deal with this situation is to grind shear onto the face of the punch, which will cause the slug to flex as it is cut, so that it springs away from the punch and down through the die.

BEN SCHNEIDER, *West Orange, NJ*

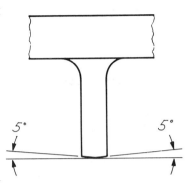

8.24 Figuring compressive stress saves diesets

All die designers know the formula to find the piercing or blanking force required for perforating work. But one should also know when that force will cause the punch or bushing to "sink" into a soft semisteel dieset when a hardened backing plate is not used. Experience shows that any vertical play in a punch will loosen it in its holder, which may result in misalignment or pullout. And any compressive stress exceeding 20,000 psi on the head of a small punch or die bushing will cause such play.

Calculating this stress is simple with the following formula, in which S is compressive stress in psi, P is the piercing force in tons, and A is the surface area in sq. in.:

$$S = \frac{2000 \, P}{A}$$

Both punch and die bushing can be figured the same way—but in the latter case be sure to subtract the area of the slug hole.

ERNEST J GOULET, *Middletown, Conn*

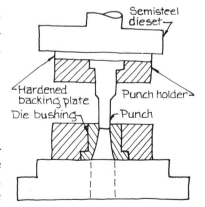

8.25 First aid for draw die

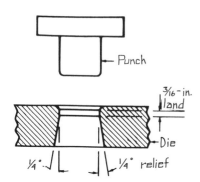

Too big a reduction in the draw die will result in premature scoring and breakout of the drawn shell. One possible fix that will save redesigning and remaking the die would be to add a 1/4° taper relief to the normally straight walls of the draw die, leaving a draw land of about 3/16 in. This reduces drawing friction and gives good results.

BEN SCHNEIDER, *West Orange, NJ*

8.26 'Flip' inserts to change bend angle of die

This die can be adapted to produce various bend angles and any arm length—equal or not—by flipping its inserts, changing the punch, and adjusting its stops.

The block-die inserts are made in standard square sizes, so each set can be rotated for the four different angles machined on the corners. If additional angles are needed, there's relatively little work to prepare another set. A variety of different-angled punches are required, of course.

The inserts are held in the die holder by a single screw in the center, which can be either a capscrew up from the bottom into a tapped hole (as illustrated) or a countersunk screw down from the top into a tapped hole in the die holder.

This particular bending die design is most useful in time and cost savings for short-run jobs.

FRED STAUDENMAIER, *Bedford, Quebec, Canada*

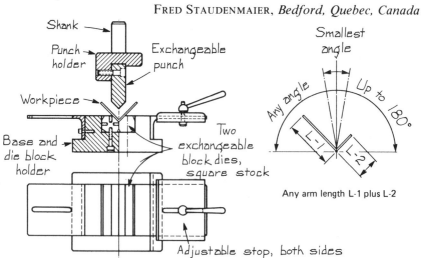

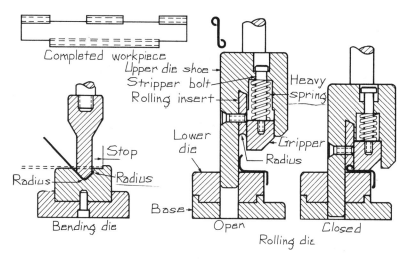

8.27 Forming double-rolled hinges

Producing double-rolled hinges is no trick at all with the two tools shown in the drawing. Not only do they work perfectly, they were also easy to make.

Following a two-step blanking and piercing operation on the workpiece, the bending die (left) is used to put two small radii and a leg on each side of the part in a two-step operation with the part flipped between them. The slight radius at the outermost edge is used to give it a better start in the rolling die. The stop point that determines the width of the leg in the bending die was found by trial-and-error.

The roughly Z-shaped part is next completed in a two-step operation in the rolling die, which is illustrated in both open and closed positions. When the press ram comes down, the hardened, spring-actuated gripper first clamps the part tightly, and the rolling insert, which is also hardened and which has a highly polished radius, then closes to complete the rolling operation that was actually started in the bending die. Again the part is simply flipped over to duplicate the operation on the other edge.

Minor adjustments to get the best diameter to fit the hinge pin are made by grinding off the bottom of the gripper and adjusting the press stroke.

FRED STAUDENMAIER, *Bedford, Quebec, Canada*

8.28 Guide plungers center strip

Plus and minus tolerances in the width of strip feedstock—whether you buy it in the "desired" width or slit it yourself—can present production problems in certain types of presswork. One such problem area is when holes have to be

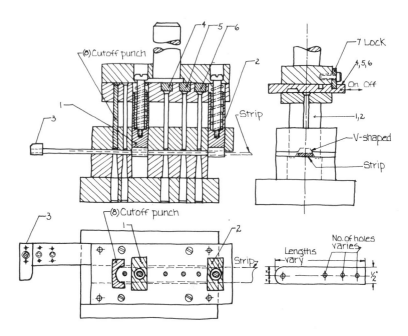

blanked precisely on the strip's center. And the variation of strip width, even though it's within tolerance, is further compounded by the need to provide clearance to facilitate free travel through the die for even the widest strip.

To solve this problem, the die shown in the drawing features two spring-loaded guide plungers (No. 1 and 2) with V-shaped cutouts at the bottom end to center the strip before piercing or blanking takes place. Despite width variations in the stock, the die produces no rejects.

Two additional features are incorporated in the die: One is to accommodate the three different lengths in which the part is required, and the other adapts it to different numbers of holes. The stock-stop (No. 3) is easily repositioned to any of three different locations to accomplish the first adaptation. To handle the variable number of holes, any or all of horizontal slides (No. 4, 5, or 6) can be repositioned to deactivate the desired needle punches. And note that both of these alterations, either part length or number of holes, can be done without removing the dies from the punch press.

FRED STAUDENMAIER, Bedford, Quebec, Canada

8.29 Hex-key/pin-punch combo

When assembling and disassembling die sections to die shoes, the most commonly used tools are Allen wrenches for screws and drive pin punches for dowel pins. These small items also are often misplaced when one is laid down somewhere on the bench and the other is being used.

That never happens to me any more since I welded the short end of a long-arm Allen wrench to the drive pin punch most often used with it, as shown in the sketch. Not only does this prevent loss on the bench, but also each serves as a handle for the other.

Further, a stand can be made for an assortment of various sizes simply by drilling a series of holes in a block of wood.

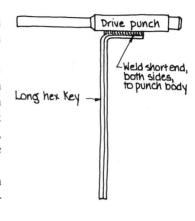

GERARD CATALDO, *Revere, Mass*

8.30 How to estimate marking pressure

Considering the press tonnage required for a progressive die that has piercing, cutting, drawing, trimming, forming, stamping, and blanking operations, how do you estimate the required stamping or marking force?

For cutting operations, I multiply the total perimeter of the cut by the stock thickness by the tensile strength of the material. For forming and drawing I do likewise—as if the part were being cut along those lines, too. But the force added by any marking operation cannot be overlooked. We tested it as follows:

We pressed a 10-character 3/16-in.-high flat stamp into soft cold-rolled steel with a hydraulic press equipped with a gage that gave us a reading of 3 tons for an impression depth of 0.004 in. A similar stamp with characters 3/32 in. high under 3 tons gave us impressions that were 0.006 in. deep. The stamp in our production die has 50 characters 9/64 in. high, so we multiplied 5 x 3 for a 15-ton estimate.

Finally, of course, any spring pressure used must also be added into the total for the ultimate tonnage.

ERNEST J GOULET, *Middletown, Conn*

8.31 How to prevent trouble with slugs

Conventional elimination of slugs and blanks downward by means of gravity is generally trouble-free, but occasionally—perhaps because of unusual conditions—special design considerations may be desirable. Here are two:

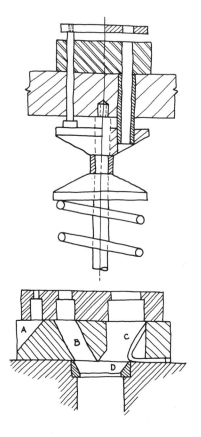

The first diagram shows the stationary portion of a small, triple blanking die of basically simple design. Its only unusual feature is the evacuation method for the blanks. To actuate the stripper, a substantial compression spring was required, the OD of which was so large that it blocked the exit holes for the blanks. To eliminate this problem, suitably sized tubes were added to extend the blank exit holes through the top of an auxiliary device that transmits the spring pressure and also ensures deflection of the blanks beyond the spring OD. The device consists of a pair of shallow cones apex-to-apex and separated by a short tube.

The second sketch shows how to avoid the problem when the situation is essentially reversed. Here the tool (a progressive die) is too large in relation to the press. In other words, the die is larger than the opening beneath it. A variety of solutions are shown:

Small slugs from preliminary stations that are too far away from the center of the press table are evacuated by depositing them on the table through angled openings in the die shoe, as at A on the drawing. To facilitate sliding of the slugs, the angle should not be flatter than 45°. When the exit hole for small slugs is near the central hole in the table, the chute can be ramped in that direction, as at B. For larger slugs or blanks, a similar arrangement can be provided, but with the addition of adequate sheetmetal chutes (C). Such chutes must be fastened securely (with screws) to the bottom of the die shoe.

Another consideration is shown at D. Most evacuation holes in bolster plates or press tables are counterbored for various plungers (for ejectors or blank-holder actuation, etc) when required. But this forms a step that may catch oiled and sticky

Forming and Press Tooling

slugs and cause problems. To prevent this, a properly tapered ring or a sheetmetal funnel should be installed, as shown at D.

FEDERICO STRASSER, *Santiago, Chile*

8.32 Lathe does rim-rolling job

In producing an experimental run of hose reels—a job that couldn't justify any great expense for tooling—we had already made some covers that looked as if they could be modified to serve as ends of the spools. Trouble was that the cover flanges weren't formed over far enough and presented a potential danger of cutting the hose that the reel was supposed to hold. Simple tooling was devised to modify the covers by rolling the rims past 90° in a lathe.

The 14-gage disk was mounted on the lathe's faceplate, and a special rolling-tool platform was bolted to the headstock and supported by a single leg below it that rested on the carriage. A manually-actuated lever was horizontally pivoted on this with a roller bearing fixed at one end as the actual rolling tool, which brings the flange position to about 90°. A second roller is mounted on the compound at 45° to form the flange well past a right angle, thus avoiding any damage to the hose.

ROBERT J PHILLIP, *Oshkosh, Wis*

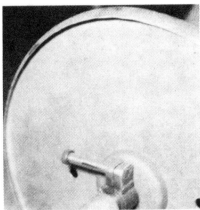

8.33 Minimize draw-punch wear

Earing material has uneven resistance to metal flow when it is drawn, producing a squarish-looking flange on the drawn part. The resulting irregular shell mouth is very

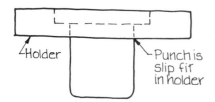

wearing to the punch at the eared areas. To distribute this wear over the entire punch surface, let the punch be a slip-fit in its holder, so that the punch will rotate slightly with each draw and present a fresh surface to the raised areas of the shell.

BEN SCHNEIDER, *West Orange, NJ*

8.34 NC punch's tooling adapted for non-NC

When the company I work for acquired its first NC punching machine, it was accompanied by a large assortment of punches and dies. To make more-effective use of this tooling, I machined the necessary punch and die holders to adapt the new tools for use on a Di-Acro hand punch that had been in mothballs for some time. Further, I designed and built a set of sliding, removable stripper plates, which are extremely valuable when setting up the Di-Acro or punching to a layout. They simply slide back so that the work can be viewed, and then slide forward for stripping the part.

The photo shows the old manual press and its accessories, which now are put to useful work for making patterns, templets, and short runs of parts.

C A SCHWANDT, *Raytown, Mo*

8.35 Nibble big holes

Seldom does the small range of dies on hand match the wide range of sheet-metal work that has to be done. So when we had to produce some 10-in.-dia holes on short notice, we designed a low-cost die that removes a section at a time.

The work pivots about a nosed pin in the center of the hole section to be removed. The method also allows the sheet metal to be turned over for easier handling during part of the operation.

FRED STAUDENMAIER, *Bedford, Quebec*

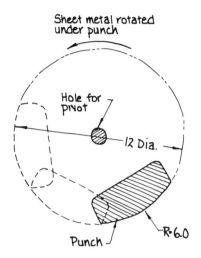

8.36 One-stroke tool cuts wire, forms both ends

The illustration depicts a tool used for the manufacture of small wire hooks; in a single stroke, it cuts off the wire stock, bends a loop in one end, and forms a 90° bend on the other.

The channel-shaped (in the top view) tool carrier is welded to the base plate

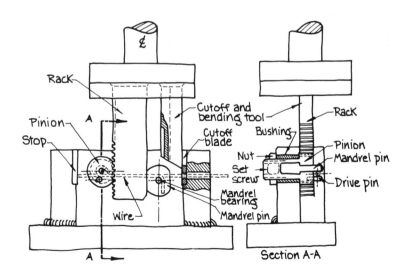

prior to the final machining and boring for mounting the pinion (in a hardened bushing) and mandrel.

The operating sequence is as follows: Unhardened wire is fed in from the right until it reaches the stop. As the upper tool descends, the rack drives the pinion, and the eccentric driving pin on the pinion forces the wire around the mandrel in the center of the pinion. As the downstroke continues, the cutoff tool shears the wire to length and then bends the wire over the right-hand mandrel to the required 90° angle.

For short- to medium-run quantities, the finished parts can be removed manually or by an air blast. For long-run operations, provision has been made for retraction of both mandrels (by cam or pneumatic action) on the upward return stroke so that the workpiece will simply drop out of the tool.

FRED STAUDENMAIER, *Bedford, Quebec*

8.37 Pencil erases die problem

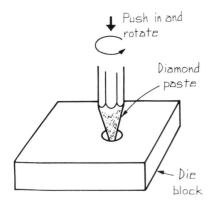

When a piercing die has been ground down, in resharpening, to below the land of the die hole, difficulties may begin to arise with slugs pulling up and creating jams or other nuisances. A cure can be effected by applying some medium diamond lapping paste to the tapered point of a wooden pencil and briefly rotating this in the die hole as illustrated.

This action creates a small negative angle in the die hole. The slug will then be wider at the point of cutting than the hole it enters, and it will not pull back.

BEN SCHNEIDER, *West Orange, NJ*

8.38 Pilot for a split stamping

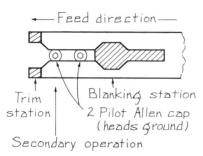

We had some trouble holding the distance between the fins of a blanked part as it passed through the die. Once the connecting web was trimmed away, the fins tended to come together—knocking the finished parts out of spec.

The fix was simple: Two flat-head 1/4-20 Allen screws were replaced with Allen capscrews, and flats were ground on the sides of the screw heads to act as internal guides along the center-line of the strip. These maintained control of the split strip's position and allowed the necessary operation to be done on fins—in spec.

BEN SCHNEIDER, *West Orange, NJ*

8.39 Plastic punch pad

The most practical backup block for punching shim stock is a piece of urethane of about 80- or 90-durometer hardness. A 4 x 4-in. piece about 1/2 in. thick is good. Just place the shim stock on the urethane block, position the punch, and rap it with a hammer. The method is especially effective and clean-cutting on very thin shim stock.

CHESTER MURDZA, *Trenton, NJ*

8.40 Progressive die boosts output, cuts scrap

This die redesign served to boost production over its predecessor by 42% and simultaneously reduced scrap by 26%. And despite the cutouts in the cross-shaped part, nine of the eleven punches are simple rectangles in section.

The progressive die has six punches in the first stage, two of which (the "A" punches) act as notching punches and also narrow the stock to correct width. When the stock is advanced, they also register it on the progression to maintain the correct feed amount. The remaining four punches in this stage (the "B" punches) blank out scrap.

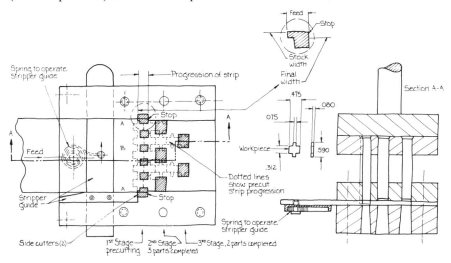

In the second stage there are three cutoff punches, the central one cutting on three sides, and the other two on two sides each. Three workpieces are completed at this stage.

In the third and final stage there are two cutoff punches, each cutting on one side only, and two more workpieces are completed here.

The moveable stripper guide, which is spring-loaded, plays an important part in holding the stock in perfect position.

As for the production rate of this tool, the die has been running at 120 strokes per minute, producing five parts per stroke, for a total rate of 36,000 units per hour.

FRED STAUDENMAIER, Bedford, Quebec, Canada

8.41 Punching from inside eases notching

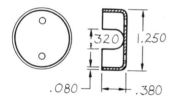

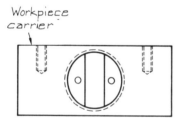

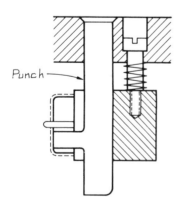

This cup-shaped part (upper left in drawing) was being produced in large volume and presented no difficulties in either drawing or punching the small holes in the bottom. But the rounded slots in the sides were being milled because use of a horn die (or a cam die) was deemed not possible. No die plate of standard design would have resisted the heat treatment, much less the cutting stresses.

To eliminate the time and cost of the milling operation, a new type of notching die was tried and proved successful. The basic idea was to punch the shell sides from the inside out. And the tool was designed and built using as many standard components as possible.

Practically standard is the die plate. The die opening is somewhat longer than is strictly necessary for proper guiding and backing-up of the heel of the notching punch. The cutting portion of the die opening has an angular clearance of 1/2° to 3/4° per side, while the guiding surfaces are straight (perpendicular to the top surface of the die plate).

The punch is quite special, composed of a rectangular body with a salient cutting portion jutting out from it. The body serves for guiding, alignment, and stripping; and the cutting point does the actual notching in combination with the cutting portion of the die opening.

The component labeled "workpiece carrier" on the drawing is a combination stripper-nest made of a rectangular piece of steel and carrying two small segments and two round dowel pins to align and hold the workpiece.

In operation, with the shell located on its nest (the segments are snug fit in the shell), the press is tripped and the ram descends. The workpiece is held snugly against its carrier by the spring-loaded holder slide as the ram continues downward to cut the first notch. After the return stroke, the workpiece is rotated on the nest 180°, and the press is tripped again to produce the second notch, which completes the part.

FEDERICO STRASSER, *Santiago, Chile*

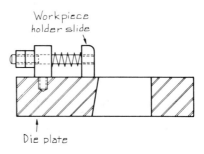

8.42 Punching in an arbor press

Every shop has occasions when it's necessary to put a few holes in light-gage stock. If proper punching equipment isn't available, the job can be done with an arbor press—provided that arrangements are made to eliminate play in the ram.

With most arbor presses, the middle and upper portions of the ram suffer most of the wear. And with the ram fully extended, there's usually a good deal of side play at the lower end just because it's so far below the housing. To prevent, or at least reduce, problems resulting from this play, we fabricated a stand, as shown in the drawing.

The top plate has a counterbored hole to accept dies, while the ram mounts the punch and a rubber stripper. A few holes drilled and tapped in the table provide the means

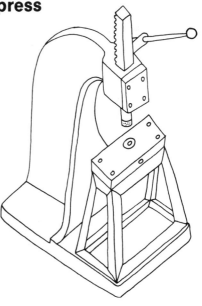

for mounting stock guides. The stand itself can be either clamped or bolted to the base of the arbor press, as desired.

CLINT MCLAUGHLIN, *Jamaica, NY*

8.43 Push or pull roll-feed handles long pitches

We had a problem in feeding strip into a power press at pitches of 350 mm (14 in.) and over. It was just not possible with conventional roll-feeds that were available. So I designed a feeding system to either pull-feed or push-feed the strip—not only for long pitches but also for short ones—and do it accurately.

The system consists of four easily made units: the roll feed, an adjustable cam fitted to the crankshaft of the press, a friction "pullback," and a spring-loaded stopper for final positioning. The accompanying schematic is largely self-explanatory.

The roll-feed is independently driven through a variable-speed drive. The top roller is spring-loaded downward onto the strip but is held away from the bottom roller by means of levers and a linkage actuated by the cam. The cam is so designed that feed is initiated by lowering the top roller onto the strip 20° after bottom dead center and ended by lifting it at top dead center.

To obtain the required pitch, it is only necessary to increase or decrease roll speed. The feed length is then set to feed a few millimeters more than the required pitch. As the strip is pulled forward, it pulls the moving carriage of the pullback unit, which uses friction pads made of felt or fiber. When the

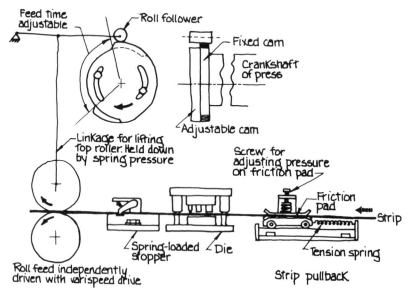

tension spring has been extended to its limit, the strip simply slips between the friction pads. Pressure on the strip can be adjusted by the pressure spring over the friction pad.

When the roll-feed releases the strip, the tension spring pulls the carriage back, along with the strip. The pawl of the spring-loaded stopper, which allows the strip to move forward freely, now locates the strip in any of the punched holes or slots in the strip skeleton. A very accurate pitch is obtained this way.

KEVIN PETERSON, Bombay, India

8.44 Realigning die pillars

After heavy use of a dieset, the die-shoe pillars may spring out of alignment—Especially after a bad jam—which results in binding between pillars and bushings and misalignment of punches and dies. And it's both expensive and time-consuming to mount the die on a new shoe.

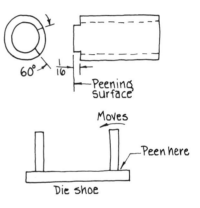

To return the pillars to squareness and parallelism, an old die-bushing can be converted into a peening tool. Leaving about 60° of the original end surface to act as a peen, grind about 1/16 in. off the end of the bushing. Then slide this bushing over the pillar to be straightened, and press this bushing into the die shoe.

Depth of pressing controls the amount of pillar movement. This can be checked with a V-block clamped to the pin, using an indicator on the upper side.

BEN SCHNEIDER, West Orange, NJ

8.45 Reinforcing a stamping

A relatively small stamping made of thin stock required a longitudinal stiffening rib, which in turn required a specially designed progressive die, illustrated by the accompanying drawings of the part itself and the strip skeleton.

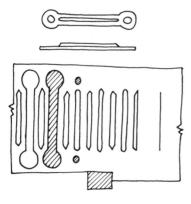

Because the stock is so thin, the rib cannot be formed by stretching, but must be produced by compound bending. The material has to come from the adjacent stock lying between two blanking openings in the strip. Since the material cannot be gained by stretching, there has to be a way to permit it to move to where it is needed.

Furthermore, the outer contour has to be blanked after forming the rib so distortions and deformation of the contour can be avoided, which is also true of the two holes at the ends. Standard stops (pins or bars) cannot be used because of the thinness of the stock, so a French stop is employed.

Finally, due to the small part size, center distances in the strip and the die were too close, so a few idle stations were incorporated in the tooling to avoid weak points between openings in the die plate.

First station—the strip is slit on both sides of the future rib, with as much clearance around the contour as possible, and with the longest slits possible.

Second station—is idle, as is every subsequent even-numbered station.

Third station—the rib is formed. Material for forming the sides of the rib is drawn from the scrap portion of the strip. There is no appreciable stretching of the stock, only bending. Distortion opens up the slits.

Fifth station—a French stop.

Seventh station—The two small holes are punched conventionally.

Ninth station—Blanking out of the finished stamping, again in a standard operation.

FEDERICO STRASSER, *Santiago, Chile*

8.46 Roughened die helps control metal flow

Under swaging pressure, metal will flow in the direction of least resistance. In some cases the swaging die cannot control, or hold, the part because the part is not entirely encased as it would be in coining, but is worked as part of a strip in a progressive die. Clamping part of the strip to lock it on die closing can cause a bent and distorted appendage on the swaged area.

By roughening part of the swaging surface of the die with a medium grade of diamond paste, you can increase the frictional resistance to metal flow. Or you can scratch lines across the flow direction with a carbide scriber. Of course, care should be taken so that this roughening or scratching is not located where it will affect the finished part, since the marks will show up on the swaged surface.

BEN SCHNEIDER, West Orange, NJ

8.47 A safe die for flat-rolling

The sketch depicts a safe and simple die design for rolling operations such as those used in the production of hinges.

The blank—with a preformed end—is put on the movable bottom member with the opposite end against the shoulder, as shown. When the press is tripped the upper member descends, simultaneously rolling the edge and pushing the lower member sideways. As the ram retracts upward, the spring-loaded bottom member returns to its rest position. As the next blank slides into position, it ejects the previously finished stamping.

The stationary heel at the left of the base serves both to determine the starting (rest) position of the bottom member's travel and to resist the lateral force of the upper member during the rolling operation.

Safety is ensured simply by adjusting the press ram stroke so that there is less than 1/4 in. of clearance between the two tool members when the ram is up. This is sufficient to feed and eject workpieces, yet is not enough to allow the operator's fingers to enter. Also, the return-spring is shielded, although this feature is not shown in the sketch.

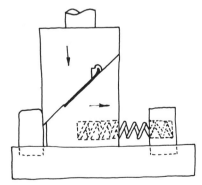

FEDERICO STRASSER, *Santiago, Chile*

8.48 Safe, simple check for punch clearance

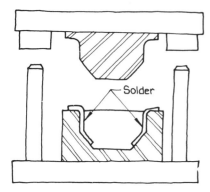

To check the location and clearance of a forming punch in its mating die, just cut a few short lengths of soft-solder wire and lay them on the inner surface of the die. Then allow the mounted forming punch to descend into the die, squeezing the solder. Upon opening the dieset, you can then "mike" the pressed solder strips and compare their thicknesses as a check on concentricity of the punch and die and on proper clearance.

BEN SCHNEIDER, *West Orange, NJ*

8.49 Second-operation die for accurate parts

In a basically simple right-angle bracket, it was necessary to punch a hole after bending so that the distance from the hole to the inside wall of the bent leg could be held within close tolerances despite rather ample tolerance on stock thickness. An almost standard hole-punching tool was built, but with some special features.

The already bent component is put on the die plate, which features a nest for lateral alignment of the stamping. When the press is tripped and the ram begins to descend, a supporting heel, spring-loaded and pivoted on the punch plate, contacts the part's leg and presses it firmly against the side of the die plate. This action, powered by a fairly strong spring, accommodates any variations in stock thickness.

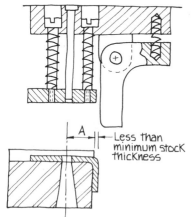

Further descent of the ram brings the spring-loaded stripper into contact with the other (horizontal) leg of the stamping and holds it firmly against the top of the die plate in such a way that the component is completely trapped and immobilized prior to actual punching. The actual punching takes place as the ram descends further.

If safety demands indirect feeding, the die plate can be mounted on a slide, and the stamping will be carried in and out by that means.

FEDERICO STRASSER, *Santiago, Chile*

8.50 Second-operation punching safely, fast

The illustrated tool—basically a standard punching die equipped with an automatic stop and ejector—has proven highly satisfactory for high-production second-operation punching. The indirect manual feed is quite safe—there's no need to reach in with hand or fingers—and the tool can be surrounded by a protective guard.

The drawing shows a tool made for punching concentric holes in washer-type stampings, but the design can be adapted to other outer shapes and even off-center holes.

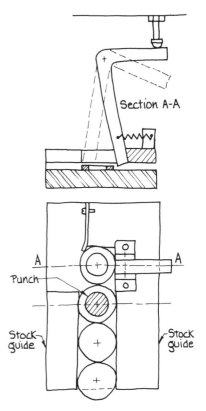

The blanks are fed into the die manually through a channel formed by the die plate, stationary stripper, and two stock guides. When the channel is full, the first blank in line is pressed against a nest formed by a shoulder in the back stock gage nd a light flat spring. This nest is located so that the second blank in the line is accurately positioned over the die for punching.

As soon as the piercing punch penetrates this second blank, ensuring its immobility, the spring-loaded ejector is actuated by an adjustable screw in the punch holder to positively eject the first blank (which remains unpunched, but which can be recycled subsequently). When the piercing punch clears on its upward stroke, part No. 2 moves into the nest to position blank No. 3, and so on.

Length of the channel should be such as to allow the last blank in line to protrude sufficiently to permit finger pressure to nest the blanks.

The design as shown is suited to short-run production, providing automatic operation with manual feeding. However, it can readily be adapted to full automation by adding a hopper and feeding slides synchronized with the press stroke.

FEDERICO STRASSER, *Santiago, Chile*

8.51 A second-operation punching setup

It's often necessary to punch holes in the bottom ends of drawn cylindrical shells, usually in a second-operation die. Such tooling is simple enough to design, build, and operate, but removal of the finished work can be a problem. It is sometimes overlooked that the ejection phase really requires two actions: (1) freeing the work from the tool, and (2) actual removal or ejection. And the two separate phases of the operation should be dealt with individually.

In the case at hand, after punching, the workpiece tends to cling to the dieplate, which also serves as a sliding-fit nest for the workpiece ID. The sketch, however, diagrams one practical solution to the problem, and the design has proven very satisfactory in practice.

The tool's design is quite standard, with a stationary dieplate/nest, a punch, and a spring-loaded stripper/blankholder. In addition, two specially shaped hooks are mounted under light spring pressure so that the points are loaded inwards.

In operation, with the drawn shell on the dieplate, as the press's ram descends (and, incidentally, a long stroke is necessary to provide clearance for introducing and removing the drawn workpieces), the hooks

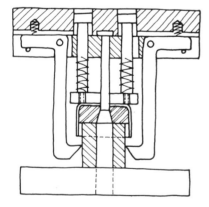

slide down the sides of the shell until they snap in underneath; then the stripper/blankholder contacts the shell bottom and presses it against the dieplate; and finally the punch pierces the shell.

As the ram ascends, the punch is first stripped from the shell bottom, then the stripper plate lifts, and finally the hooks lift the shell from its dieplate/nest. At the top of the stroke, the workpiece is completely free, and it can be removed, say, by a synchronized blast of compressed air.

FEDERICO STRASSER, *Santiago, Chile*

8.52 See-through stripper simplifies punching

Our shop uses a number of dies that require shimming after grinding. The bench fixture shown in the sketch, used with a set of transfer punches, speeds and simplifies the job of punching the required holes in the shim stock.

After scribing the size and location of holes on the shim stock, it can be quickly positioned in the fixture because of the visibility afforded by the transparent plastic "stripper." Drill blanks, dowel pins, reamer blanks, or hardened drill rod will serve as punches.

ADRIEL L SMITH, *Belmont, Miss*

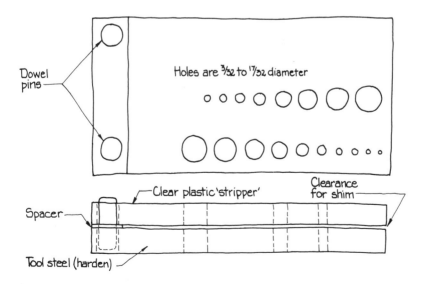

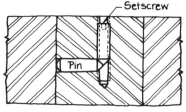

8.53 Setscrew turns corner

A hardened tool-steel insert had to be secured in a matching opening in a hardened tool-steel plate with all surfaces flush and no protruding parts.

This was done in the following manner: The hole was machined in the plate with smooth, parallel walls perpendicular to the surface. The side walls of the insert were similarly machined to a snug sliding fit. Before hardening the insert, two intersecting cross holes were drilled; one of these was tapped for a hardened, cone-point (90°) setscrew, and the other reamed to fit a hardened pin with one rounded end and the other conical.

For assembly, the pin is first inserted in its hole with the conical end inward and the setscrew just started in its hole. Then the insert is put into its hole, and the setscrew is tightened (disappearing below the surface) to drive the pin outward for clamping. For comparatively light loads, this fastening method works excellently.

FEDERICO STRASSER, *Santiago, Chile*

8.54 Setting die clearance with a nylon stocking

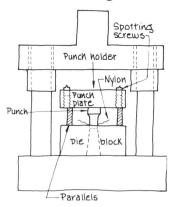

Alignment of punches and perforators in their die openings and maintaining proper clearance between punch and die are important for cleanly blanked edges. Placing a piece of nylon stocking over the die opening and gently forcing the punch into the opening—the stocking should not be cut—will provide an equal amount of clearance all around. And in cases of larger clearances, heavier pieces of cloth can be used.

To finish the job, spotting screws are

inserted in the punch plate, and suitably sized parallels are placed between the die block and the punch plate. Then the punch holder is forced down on the points to show screw locations for drilling. After the punch plate and holders are screwed together, clearance can be double checked with a die light and dowel pins can be added for permanent location.

WILLIAM HITCHEN, *Chicago*

8.55 Single die does two operations for seaming

Seaming the edge of sheetmetal usually takes two setups with two separate pieces of equipment. The die arrangement shown here, however, was designed to perform both steps in two operations of the press: In the first operation (left), the material is bent to 135° from the horizontal (the die is shown at the mid-position of its upward, return, stroke). The part is then relocated manually on top of the stripper plate for the second operation (right), in which the downstroke finishes the seaming job.

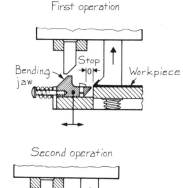

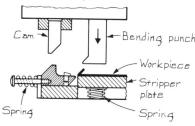

Here's how it works. During the first downstroke, the bending punch holds the work firmly against the spring-loaded stripper plate, and the material bends to 90° as the bending punch pushes it past the bending jaw. Then the cam moves the bending jaw to the right so that it bends the material to 135° over the toe of the bending punch. The die then returns to its starting position, and the bending jaw is retracted by its spring. The second downstroke then simply flattens the work to complete the seam.

FRED STAUDENMAIER, *Bedford, Quebec, Canada*

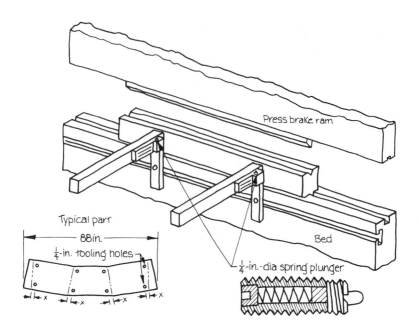

8.56 Spring plungers align part for forming

Long workpieces, particularly those with non-parallel bends to be made, can be difficult to form consistently on a press brake with conventional stops.

On some parts for which the flat blank is nibbled to size on a nibble/punch machine, we add two 1/4-in. tooling holes for each bend. These holes serve as alignment holes for the forming operation. Two 1/4-in. spring plungers are used, as shown in the drawing, to fit into these tooling holes to align the bend line with the center line of the press brake punch and die.

The technique allows the part to be formed consistently each time without requiring multiple setups or stops. And setup time is also reduced.

Gary R Gregg, Shippensburg, Pa

8.57 Spring-loaded guide handles heavy stock

For most applications, commercially available stock pushers are quite suitable for use on our dies. There are times, however, when we run heavier stock sizes and the material (which comes in 12-ft lengths) acts as a lever and renders the pushers useless. To prevent this problem, I designed a special device that, by means of a predetermined spring load, applies the desired force against the stock directly in line with the back gage of the die (see drawing).

During the press downstroke, the pivoting wedge initially contacts the

Forming and Press Tooling

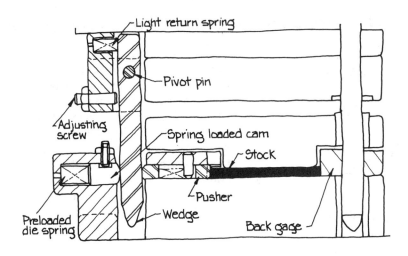

springloaded cam, which is backed up by a preloaded die spring of the desired specification. This spring offsets the small upper return spring and causes the wedge to be driven against the pusher. When the stock is fully registered against the back gage, the wedge is forced backward against the cam causing the spring to exert a force sufficient to guarantee that the stock is positioned against the back gage throughout the remainder of the press stroke.

ROBERT D WISNIEWSKI, *South Milwaukee, Wis*

8.58 Stepped arbor preforms rings and springs

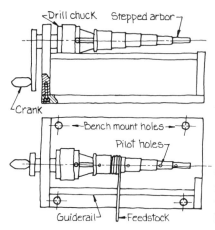

In our shop we frequently use preformed solder rings of various sizes, and we've found the illustrated gadget extremely useful when such operations are necessary. It's simply made, as the drawing shows, and the stepped arbor is turned to provide us with the most frequently used ring diameters. Rotation is by manually turning the crank with one hand, while the other hand controls the feedstock going across the guide rail, which offers a degree of control of ring spacing. When we need rings that don't match any of the arbor steps, we just chuck up a piece of appropriate barstock or a drill blank. And the device is also useful for making a range of coil springs.

ROBERT F BOWEN, *Burlington, Mass*

8.59 Stock guide oils work

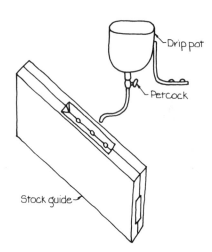

Sometimes the strip to be stamped and formed in a Multi-Slide machine needs more lubricant than the standard felt-pad wipers leave on the stock surface. If this is the case, just grind a slight angle in the inner, upper edge of the stock guide to form a shallow, V-shaped reservoir, and then grind two or three vertical channels from this to allow the oil to seep down to the strip. The oil supply is a simple drip pot positioned over the stock guide with a petcock to meter the oil that drips into the guide reservoir.

BEN SCHNEIDER, *West Orange, NJ*

8.60 Styrene test blanks

Our shop produces many short-run sheet metal panels on an NC turret punch press—some of them with as many as 300 hits. To verify the tapes, we use 0.060-in. thick high-impact styrene sheet. The obvious errors can be spotted immediately, and the material holds dimensions closely enough for inspection of all holes and cutouts.

Any errors in a styrene test panel can be covered with masking tape, and the same piece can be run again after the program has been edited. Before switching to styrene for tape-proofing, the actual work material was used, and any errors had to be welded if the material was not scrapped immediately. And the styrene costs only about 10% as much as the actual material.

Russ Brown, Ormond Beach, Fla

8.61 'Tab stop' improves on 'French stop'

The time-honored "French stop" is widely used because it can be built into the first station of a progressive die to serve all material stopping functions through the die. The new "tab stop" illustrated here will do all of this, and it also eliminates the need to increase the width of the strip that is inherent in the French stop. French stops also can involve considerable machining of die block, die shoe, and lengthy special punches. The tab stop does its job with standard square punches and die buttons that can be purchased off-the-shelf.

With a French stop, it's necessary to shear out an area along the edge of the stock that is equal in length to the full distance between die stations. With this idea, all that's required is a pair of square punches and die buttons spaced the desired distance apart. And the slugs can simply be dropped through drilled holes in the die shoe.

Here's how it's done: Mount the two identically sized square punches and die buttons at spaces equal to the required distance. Then grind away the side of the first punch (see drawing) equal to the thickness of the stock and radius

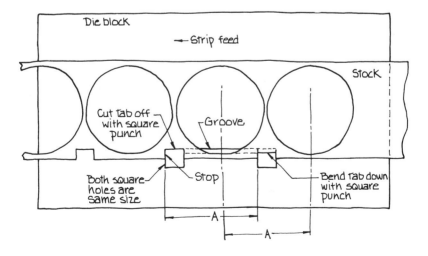

the edge so the punch will shear three sides and bend down a tab in the scrap portion of the strip. This tab slides through a slot, or groove, as illustrated, to be stopped by the far side of the second square hole. The second punch then removes the tab to allow the stock to feed—and be stopped by the next tab.

BALLARD E LONG, *Oak Ridge, Tenn*

8.62 Trim square boxes in one press stroke

Square drawn boxes are always ragged and irregular on the top edges, so they are usually made higher than the nominal dimension and trimmed afterward. This is most commonly done from the corners with a vertically-acting V-shaped blade and a horn die and takes either two strokes or four strokes of the press with intermediate repositioning of the box. Not only does this cost extra time, it often results in inaccuracies or mismatches in the sheared edges because of the need to reposition the workpiece on the horn die.

The drawing shows a tool design that is still relatively simple and inexpensive—yet allows trimming of the workpiece in one setting and one press stroke. The stationary mandrel (horn) is much the same as with the conventional tooling described above. But two V-shaped trimming blades act horizontally—one at a time—actuated by vertically moving cams, one of which allows retraction of its blade before the other blade starts to move.

The three cutting members, the blades and the horn, are made of tool steel and hardened, of course, and the blades are mounted in properly guided horizontal slides, which are automatically retracted by tension springs. This outward travel is limited by simple stops. The workpiece is removed after trimming by means of a hand-actuated knockout that is normally held in its rest position by a small compression spring.

FEDERICO STRASSER, *Santiago, Chile*

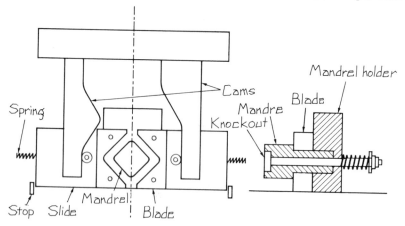

8.63 When a punch is undersize

As sometimes unfortunately happens, the tool grinder may grind off too much material from a punch so that it doesn't provide proper clearance in its mating die. Or the die may have been resharpened so often that it's down into its draft clearance. In either case, the result would be excessive burrs on the stampings being produced.

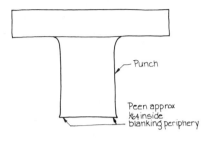

If the stamping is in thin material (under 0.010 in.), it's possible to semi-anneal the punch to about 45 Rc, then peen out the cutting edge, and shear the punch into the die to a depth below the normal shut height of the die at closing. The punch will thus be sharpened by the die, and it will hold up surprisingly well in production. It will give a clean and burr-free edge to the stamping, and it can be repeened again and again for resharpening.

BEN SCHNEIDER, *W Orange, NJ*

8.64 When trigger stops don't work

Trigger stops are usually used after the last (blanking) station in progressive dies, as shown at A in the drawing. There are cases, however, when other arrangements are preferable—primarily when stock is too thin or too thick.

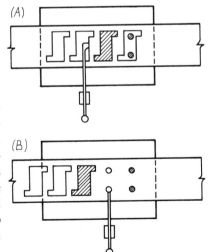

In the first case, trouble can arise because the scrap-allowance bridges between two adjacent blanking openings in the strip skeleton (the separations) are too weak and tend to deform under feed pressure. Solution to this problem in a progressive die is to actuate a trigger stop at an intermediate station with some preliminary pierced opening, as at B. At this point the strip skeleton is strong enough to take even rough handling.

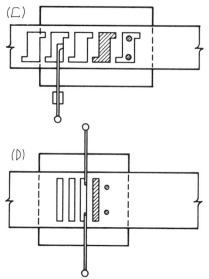

When stock is heavy, say, over 1/8 in. thick, trigger stops may be troublesome because of crowding by metal flow. In such cases it may be advantageous to locate the trigger stop so it engages the second opening in the skeleton after the blanking station, as at C.

A third situation is when blanks are long and slender. In this case, two trigger stops may be better than one (D).

FEDERICO STRASSER, *Santiago, Chile*

9 Bench Work

9.01 Add a second socket

When changing the jaws on a large vise, you never seem to be able to get more than a half turn of the Allen wrench before something interferes, making removal of the screw almost endless.

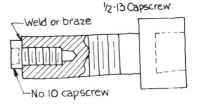

I simply drilled and tapped the other end of the capscrews to accept smaller capscrews. These were inserted and then brazed in place, although a drop or two of Loctite would probably work as well. Now, with the smaller socket on the other end, it's a quick job with a smaller key to speed the capscrews in and out—using the larger key to break them loose or set them, of course.

DAVE FERDINAND, *Chatham, NJ*

9.02 Backward hand stamping is neater

It's tedious and difficult to produce a neat lettering job with metal stamps. Here's a technique that makes it easier:

First, write out the "message" on a piece of paper. Place the proper stamp

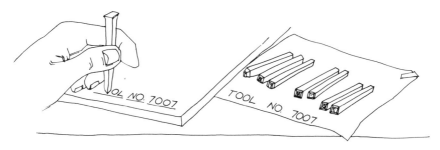

on each number or letter, working from the *end* of the message. Then draw a pencil line on the part to be stamped (to keep the lettering aligned).

Begin by stamping from right to left. This gives you a clear view as your left hand holding the stamp does not obscure the previously stamped characters. If the letter or number is to be used again in the same message, put it back on the paper in the next position it will be used.

Neat stamping requires skill and care. This technique makes it easier to do a good job.

ART DRUMMOND, Walworth, NY

9.03 Bench 'fence' adjusts

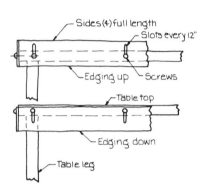

When working with small parts on a shop bench it's a good idea to have a raised edging around the bench or table to prevent parts from rolling off onto the floor. And at other times it's preferable to work without the edging. The system sketched here accommodates both conditions: the slotted edge strips can be raised when needed, and dropped away when they aren't wanted.

WILLIAM SLAMER, Menomonee Falls, Wis

9.04 Carbide tool deburrs slots, keyways

This simple tool, which can be made in a few minutes, is quite useful for deburring slots, keyways, grooves, and similar features on a variety of workpieces. Because cutting is done with the edge of the insert, almost any square or triangular carbide insert can be used, provided it has a clamping hole.

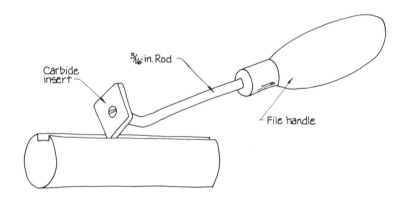

Bench Work

Depending on the size of the insert hole, a 6-32 or 8-32 hole is drilled and tapped in one end of a 6-in. length of 5/16-in. rod, and the opposite end is threaded for a file handle. The tapped end is then bent about 30° approximately 1/2 in. from the end.

WILLFRED G MOORE, *Chicago, Ill*

9.05 Differential threads set adjustable spacer

It's possible, of course, to make spacer rods of adjustable length (for clamping setups) by using a right-hand thread on one end and a left-hand thread on the other, as in a standard turnbuckle. But it's also possible, and sometimes more convenient, to use two right-hand threads of different pitch. It usually depends on the taps and dies you happen to have available. The sketch is self explanatory. By turning the "nut" in one direction, the threaded rods will expand or contract, depending on direction.

ANTHONY J MORRIS, *Van Nuys, Calif*

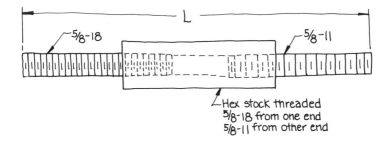

9.06 Disks deburr small holes

It took me about 20 min to make six small disks of various diameters from drill rod—hardened and ground to about 0.020-in. thickness.

Why? Chucked in the cross-slotted collet of a pin vise, they make excellent deburring tools for small holes drilled in brass, copper, steel, aluminum, etc. They work extremely well in plastics.

Most of the time I use these deburring disks by hand, but, in some cases, I have put the pin vise in a lathe or mill and used it under power.

RON STANWICK, *Englishtown, NJ*

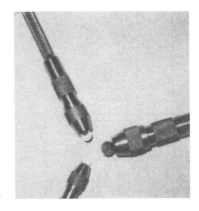

9.07 Handy chip blower clears blind holes

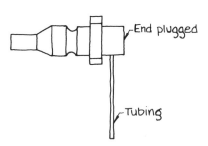

A handy device for blowing chips out of blind holes—and it works to depths of at least 6 in.—is a length of small-diameter thin-wall tubing, preferably metal, soldered or epoxied into a plugged air-hose fitting of whatever size and type is used in your shop. The end can be plugged with a pipe plug or a drive-in plug.

DAVID R CARLSON, *Manchester, NH*

9.08 A handy clamp borrowed from jewelers

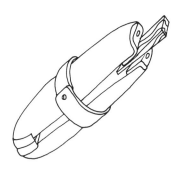

This small hand clamp, with a wooden body and metal inserts and a metal band around it, may be the ancestor of all vises. It's used by jewelry makers, who use leather inserts and band, to hold rings and other small adornments they're working on, but I've never seen one in a metalworking plant.

A wooden wedge pushed in at one end closes the other end to clamp the small workpiece. It's suitable for holding work for soldering, grinding, or light filing. Body size and shape are selected to fit the user's hand.

MAURICE CONKLIN, *Mississauga, Ont, Canada*

9.09 Holding thin parallels

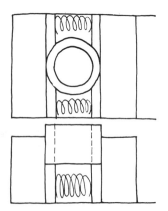

Sometimes a thin-walled workpiece must be set up in a machine vise with thin parallels, which display an annoying tendency to topple out of position. The problem can be solved by supporting the parallels with some compressible medium—a pair of light compression springs (as shown), two rubber blocks cut from erasers, or even balled-up newspaper in an emergency.

FEDERICO STRASSER, *Santiago, Chile*

9.10 Make a carbide scraper

Here's a tool that's ideal for scraping off old gaskets, paint, and rust from flat surfaces. Just take a carbide turning insert and silver solder it to the end of a screwdriver that fits your hand comfortably. Then use it much as you would a putty knife and it won't scratch or gouge the work surface.

VINCENT CHENEY, *Lyme, NH*

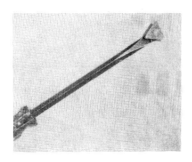

9.11 Make it a 'pull' saw

Nine times out of ten the hacksaw blade you use in your keyhole-type sawblade holder will bend or break before it wears out. Yet there are many times when it's a handy tool because you can't get a standard hacksaw frame to do the job because there just isn't clearance for it. What you do is reverse the blade in the holder so that it cuts on the pull stroke. That puts the blade in tension, and it doesn't need rigidity to keep from bending and breaking.

BEN SCHNEIDER, *West Orange, NJ*

9.12 Modification makes deburring tool handier

When using a rotary deburring tool on parts with square corners, the freely rotating blade is sometimes a bit difficult to position properly for starting the corner cut. To alleviate this problem, just modify the tool by removing the blade and inserting a small compression spring as shown in the sketch.

This spring puts a light axial load on the blade so that it stays in place at the edge of the work, yet still permits the blade to swivel as necessary during use. This simple modification makes the rotary deburring tool easier to use for all types of work.

ALVIN ELBERT, *Portland, Ore*

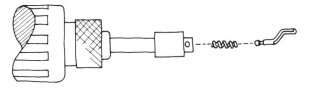

9.13 Parallel clamp improved

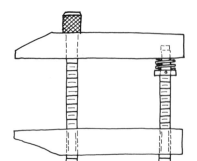

When using an ordinary parallel clamp, it is not easy to tap the clamped pieces into position because the clamp is always either too tight or too loose. To eliminate this problem, we added a spring to the clamp, as shown in the drawing.

Now the clamp can be snugged up on the parts, while still leaving them movable for final alignment. When position is correct, the rear screw is tightened further so that it bottoms in the blind hole and functions like a normal clamp.

Modification only requires addition of the collar and a spring.

CLINT McLAUGHLIN, *Jamaica, NY*

9.14 Pin modification lets vise grip odd parts

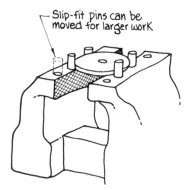

Slip-fit pins can be moved for larger work

You can't call it a "pin vise" because that name's already been taken, but a series of holes drilled into the tops of both jaws to accept slip-fit pins enhances a vise's capability for holding disks, odd-shaped parts, and thin work. Reason for the slip fit is so the pins can be repositioned for different shapes. And the technique will work for either a bench vise, a drill-press vise, or even a milling-machine vise.

JOSEPH HORNIS, *Lorain, Ohio*

9.15 Plastic belting a handy polishing tool

The polishing of small carbide tooling components can sometimes be a time-consuming job. I've had a lot of success by doing this with diamond paste and a pieces of 3/16-in. urethane belting chucked in a high-speed hand grinder. It's the type of belting used in some conveyor drives.

I put a piece of the belting in the grinder with about ¼ in. sticking out, and I use only the face of the belting. The diamond paste is applied to the surface to be polished, and the urethane seems to hold the paste in the working area, allowing it to lap the surface. Other applicators seem to push the paste aside, taking longer to do the job and not doing it as well. And when the tip of the belting gets glazed, just trim off a bit of the end.

ANTHONY F SIRACUSA, *Sewell, NJ*

9.16 Punches don't fly from non-skid hammer

When using steel stamping punches, an effective way to keep the stamps from flying away from the hammer is to tap the hammer a few times on a strip of 60- or 80-grit emery cloth that has been laid on a hard, flat surface. Some granules of grit from the emery cloth will be retained on the face of the hammer, making it a skid-resistant surface. Then the stamp or punch will not fly away, even though it may not be struck a perfectly square blow with the hammer.

MARTIN J MACKEY, *Parma, Ohio*

9.17 Setup plate fits mill vise, bench vise, lies flat

This simple setup plate has proved to be very useful—especially for prototype and short-run machining—in enhancing the utility of a machine vise on a small, vertical mill. Work can be clamped to it easily or, in the case of plastics, can be held with double-coated tape.

The retractable edge rail, which fits either edge, is useful for lining up some workpieces. And the 3/4-in.-sq bar in the bottom, which is counterbored underneath for Allen-head screws attaching it to the plate, can be clamped either in the milling-machine vise or in a bench vise, or can be removed to allow the plate to lie flat on the bench during layout of parts clamped on it.

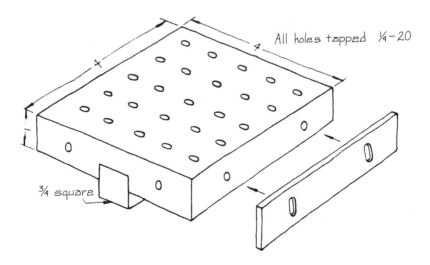

The plate also facilitates transfer of work from one machine to another—mill, drillpress, grinder, even a lathe with a four-jaw chuck—without unclamping; there's always something to hold onto.

Although any metal could be used for the plate, mine is made of 2024 aluminum so that additional details can be machined easily if necessary to accommodate some workpieces.

RON STANWICK, *Englishtown, NJ*

9.18 Simple grabber pulls blind dowel pins

Use of blind dowel pins is poor design and should be avoided whenever possible. But it just isn't always possible, so you must foresee some method for pulling them out for maintenance or repair.

The sketch shows one solution to the problem: a kind of parallel clamp with one bar extended and bent into a U-shape to provide a driving surface. To use the tool, the exposed portion of the pin is clamped tightly, and a couple of hammer blows at the outboard end will quickly pull the dowel.

FEDERICO STRASSER, *Santiago, Chile*

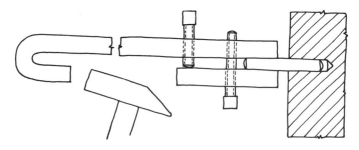

9.19 Storing parallel clamps

We used to keep parallel clamps in a drawer, where they always seemed to end up in a jumble. We've now simplified the job of getting at the clamps when necessary, gained back the drawer space, and still present a neat appearance. All this by mounting simple aluminum hangers on a sheet of plywood and mounting that, in turn, on the side of a cabinet.

MARTIN BERMAN, *Science Machine Shop, Brooklyn College, Brooklyn, NY*

9.20 X marks transfer spot

Chances are you already have this simple and inexpensive transfer punch that can locate the center of holes of any depth or number. A new twist drill about 0.005–0.010 in. larger than the bolt can be inserted in the clearance hole, gently tapped, then turned a quarter turn, and tapped again. A "cross hair" (+) impression will be left by the two impacts of the drill's chisel edge. Just center-punch at the cross, then center-drill. Larger holes will come out very accurately.

MARTIN J MACKEY, *Parma, Ohio*

10 Layout, Inspection, and Measurement

10.01 Adjustable angle fits height gage

A universal bevel protractor is the conventional instrument for use when layout or inspection operations call for some angle other than 90° or 180°. But that very useful tool is often too big for convenience. For these tasks I use the simple adjustable angle illustrated here, which fits in a height gage.

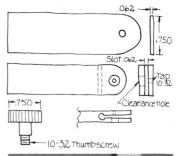

The drawing is completely self-explanatory except that it should be mentioned that by making the blade an even 4 in. long from the center of its pivot hole, an offset of 0.070 in. at the end will equal 1°. This facilitates the use of gage blocks and pins for setting angles from either a horizontal surface or the vertical blade of a try-square. Angle blocks can also be used for setting, of course, or you can set it up with the bevel protractor that you're not using otherwise.

RON STANWICK, *Englishtown, NJ*

10.02 Angle plates make V-block

About the only way to check the concentricity of parts that haven't been turned between centers is by using V-blocks. Lacking V-blocks of a size suitable for a particular part, I built one from three angle plates, as shown in the sketch.

The bottom of the base angle plate was milled flat with the part held in a standard

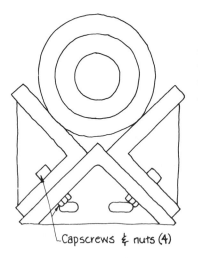

V-block. Also, two slots were milled in the reinforcing web of this angle plate to accommodate strap clamps if they should ever be needed for a setup. Finally, four close-fitting holes were drilled for capscrews and nuts to hold the angle plates together.

NILS G BRADLEY, *Westerly, RI*

10.03 Bent-rod dial-indicator holder is versatile

A few inches of 3/16-in. rod and a few minutes of work bending it to the shape shown will produce a very inexpensive and yet quite handy holder for dial indicators such as Starrett's Last Word. Used in conjunction with the standard body clamp, it can be mounted in a machine spindle, on a surface gage, or virtually any other place desired.

RON STANWICK, *Somerville, NJ*

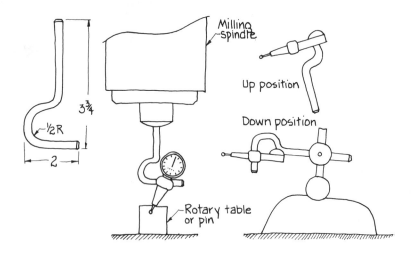

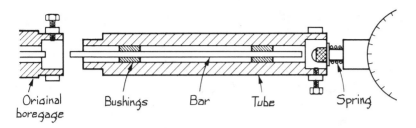

10.04 Bore gage checks 15-in.-deep holes

We had to drill some holes with a 1-in. diameter that had a tolerance of 0.010 in. at a depth of 15 in. Our standard bore gage was far too short, and we didn't have much success with go/not-go cylinder gages either. So we made an extension for the bore gage.

A length of drawn tubing was faced off on both ends and one end was turned to fit the ID of the indicator seat in the bore gage and the other end was bored to fit the indicator stem. Two phosphor bronze bushings were pressed into the tube, and a sliding fit bar was inserted. A reinforcing spring was added to the sliding shaft of the dial indicator to overcome the added weight of the long bar.

The various parts were assembled with the indicator contacting the internal bar, and the extended gage was calibrated with the help of a micrometer.

FRANZ UNGER, *Kirjat-Mozkin, Israel*

10.05 Chamfer vs radius

In many cases, a chamfer and a radius are functionally interchangeable, and the former may be a lot easier to produce. I've found the following equivalent table very handy over the years in both the shop and the drafting room.

Radius	XXX	Radius	XXX
0.016	0.009	0.375	0.220
0.031	0.018	0.406	0.238
0.047	0.027	0.438	0.256
0.062	0.036	0.469	0.275
0.078	0.045	0.500	0.293
0.094	0.055	0.562	0.329
0.125	0.073	0.625	0.366
0.156	0.091	0.688	0.403
0.188	0.110	0.750	0.439
0.219	0.128	0.812	0.476
0.250	0.146	0.875	0.512
0.281	0.165	0.938	0.549
0.312	0.182	1.000	0.586
0.344	0.201		

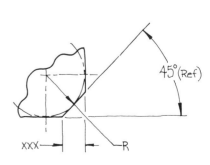

To build a more complete table, you can fill in other values by using the formula: XXX = R x 0.586.

BERNIE WAAG, *Levittown, NY*

10.06 Checking a small groove in a small bore

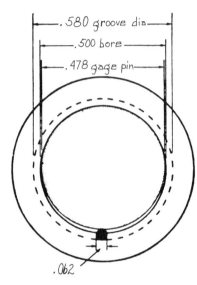

It's extremely difficult to get an accurate reading of the major diameter of a small groove inside a small bore. Here's a method that's relatively quick and simple for finding that elusive major internal groove diameter to an accuracy within ±0.001 in.

First of all, the bore diameter must be known, but this is pretty easy to measure. Then you'll need a steel bar with a diameter that's smaller than the width of the groove but larger than its depth. Insert the steel ball in the groove. then, using a set of gage pins, find the largest pin that will pass by the steel ball and wedge itself against the opposite wall of the bore. Record the diameter of the pin, add the diameter of the ball, and subtract the diameter of the bore. This will give you the depth of the groove to within 0.001 in.

Putting this into formulas, if A = ball dia, B = bore dia, and C = gage pin dia, then:

Groove depth = A + C − B, and

Major groove dia = 2 (A + C − B) + B, or 2 (A + C) − B

In the example shown in the sketch, groove depth is 0.062 + 0.478 − 0.500 = 0.040 in., and the major diameter is 2 x 0.040 + 0.500 = 0.580 in.

ALEXANDER KASS, *Jamesville, NY*

10.07 Checking the depth of internal grooves

The problem was to accurately check the diameter of grooves inside of a 2-in. bore on a lathe. Measuring width was no problem, but it just wasn't possible to gage the

groove ID with conventional measuring instruments, such as an inside micrometer.
So a simple disk was made that was a loose fit in the width of the grooves. Disk diameter was half the difference between diameters of the groove and the bore. Edges were chamfered to ensure that the disk would seat properly on the bottom of the groove. Insertion and removal of the disk was done with tweezers.

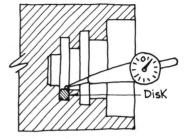

Actual measurements were made with a lever-type dial indicator clamped in the lathe's toolholder by comparing the lowest point in the bore with the highest point on the disk.

H P POHUJA, *Bombay, India*

10.08 Circle scriber needs no center

Many workpieces—rings or valve castings, for example—feature flanges that require layout and drilling of bolt-hole circles, yet have no metal in the center

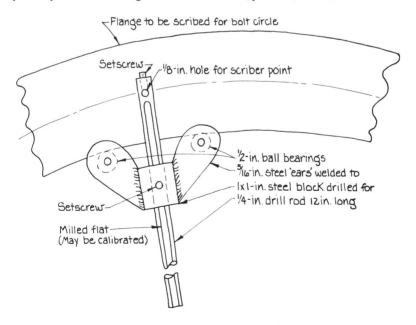

for a punch mark that would permit the use of dividers for scribing the circle. The simple tool shown in the illustration does the circle-scribing job without any need for a center; it just rides around the ID on ball bearings. Alternately, it works just as well riding around the OD of the flange, if that's preferred. And it can be used to scribe lines parallel to the edge of rectangular work, or any other shape that presents itself in the shop.

Dimensions can be varied widely, although the distance between the centers of the ball bearings on mine is about 2 1/2 in. Also, the milled flat on the length of drill rod can be calibrated with marks at frequently used distances.

GARY S MINGUS, *Louisville, Ky*

10.09   Dial caliper sets combination square

A combination square can be set with enhanced accuracy simply by using a dial caliper as shown in the sketch. Just set the dial caliper at the desired reading and tighten the lock screw. Then butt the end of the combination square's scale against the step of the caliper's sliding jaw and slide the square head into contact with the outer end of the caliper, and finally tighten the lock screw of the square.

Obviously this method only works with dial calipers incorporating the so-called four-way measuring feature, but calipers without this feature can provide the same results by using the depth rod.

The technique can also be reversed: that is, a measurement can be taken with the combination square and then can be precisely determined by reversing the above procedure.

BRUCE JONES, *Sacramento, Calif*

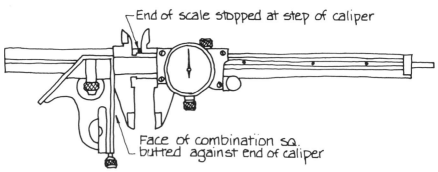

10.10   Dial indicator checks automatic toolchanger

First-piece inspection often means expensive downtime on costly NC machining centers. Continued production of that particular part can be risky until the first one's checked out. And you generally can't switch to another part because the fixturing (and often the cutting tool settings) are part of what's being checked in that first-piece inspection. Also, especially with castings, it's difficult at best to get a workpiece back in the holding fixture exactly the same for corrective machining after removing it for inspection, so you're also risking what may already be a valuable workpiece.

In many cases a simple and effective solution to this problem is to mount a standard dial indicator in a toolholder and use the NC machine itself to check out its own work. Depending on the nature of the machined surfaces to be checked, either a plunger type or a "Last Word" type of indicator may be used.

For checking the depth of several end-milled surfaces, a typical task, a plunger-type dial indicator (preferably with a 0.100-in. to 0.200-in. range) should be preset to a known dimension in your tool-setting setup. Then ask the part-programmer to prepare a program that will simply position the indicator over the milled surfaces to be checked. Using a bore-and-dwell cycle with a dwell time long enough for jotting down each indicator reading, run it through the positioning program and advance the indicator to each of its nominal settings. This provides all the data needed to correct tool settings for finishing the job in the fixture and subsequent production.

Variations for different machines, tooling practices, and workpieces will be pretty obvious, but in any case the inspection will probably take less time than removing the workpiece from the fixture. Costs are virtually nil because there's bound to be a spare indicator and toolholder around, setting time is brief, and the "check tape" is quick to program.

The check tapes should, of course, be properly labeled and kept separate from machining tapes. Also the indicator setting information and check tape number should be part of the regular setup sheet for any given job.

JOSEPH D JUHASZ, *Horsham, Pa*

10.11 Figuring countersink depth

I've been a machinist for many years, and I have always wanted a way to determine how far to advance a countersink in order to obtain the correct diameter at the surface, which is the figure usually given on the blueprint. I recently figured out a formula that does the job for any countersink angle.

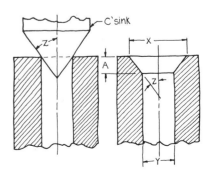

Labeling the advance depth 'A,' the formula is:

$$A = \tfrac{1}{2}(x - y)\text{cotangent } z$$

in which 'x' is the specified surface diameter of the countersink, 'y' is the diameter of the straight hole where the countersink first makes contact, and 'z' is half the included angle of the countersink. The drawing will help clarify this.

WILLARD W ANDERSON, *Pacific Grove, Calif*

10.12 Figuring screw diameter

Often you have a need to know the outside diameter of a specific screw, but a screw-size chart just isn't available. If it's a numbered machine screw, you simply multiply the number of the screw times 13 and add 60 to get the OD in thousandths. Examples:

1-72: 1 x 13 + 60 = 73 (0.073 in.)
4-40: 4 x 13 + 60 = 112 (0.112 in.)
10-32: 10 x 13 + 60 = 190 (0.190 in.)

Note that this formula is invaluable when used in conjunction with John E. Whittle's "Mental tap-drill chart," in which tap-drill size is determined by subtracting pitch from OD.

DAVID A ACKROYD, *Newport Beach, Calif*

10.13 Four V-blocks in one

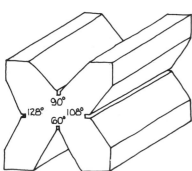

We recently had some work that required V-blocks with four different included angles for inspection of various features—thus requiring a total of eight V-blocks. To meet these multiple needs, however, I designed a single pair of V-blocks with a differently angled V on each face—two blocks for eight. Size was 2 3/4 in. x 2 3/4 in. x 2 1/2 in.

RUDOLPH CHODIL, *Norridge, Ill*

10.14 Gadget finds drill size and center location

Among the various methods for cutting special contours in steel dies, one is to scribe the contour and then proceed to drill out the unwanted material with a series of holes. One problem with this technique, however, is in selecting the drill size that best matches a contour at any particular point and then locating the center accurately.

Layout, Inspection, and Measurement

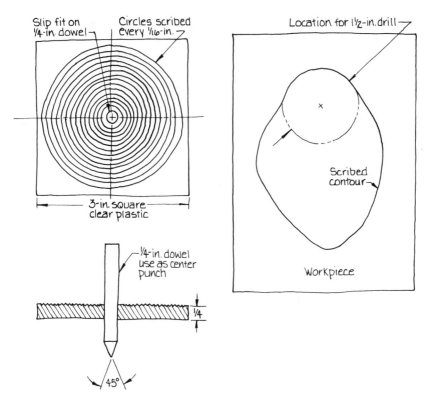

The simple device illustrated here allows just that. Just place it over the scribed line and move it around until you get a "best fit." Then punch the center point. The result allows use of the largest possible drill (so fewer operations are needed), and it gets closer to the desired contour (so that finishing is minimized).

FRANK L PELLIZZARI, *Racine, Wis*

10.15 Gage checks straightness

Checking a small (1/8 in. dia, 2 in. long) steel shaft for runout is fairly routine for an inspector using V-rests and a dial indicator. But when there are hundreds to be checked to a couple of "tenths" tolerance, the inspection time drags into excessive amounts.

After considering several possibilities, we devised a faster method—with a very simple

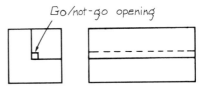

fixture that can be made to extreme accuracy and provides the advantage of four precision-ground contact surfaces.

The fixture consists of two attached hardened steel blocks slightly longer than the shafts to be checked. One block has an inside corner ground out to accept the other, and the second one has an inside corner ground out to accept the shaft diameter plus the permissible runout tolerance—no more. Acceptable shafts must pass through the fixture without pressure; any interference indicates a reject. And it doesn't require a skilled inspector.

ROBERT J MCMASTER, *McCord Winn Div, Winchester, Mass*

10.16 Gage measures countersink diameters

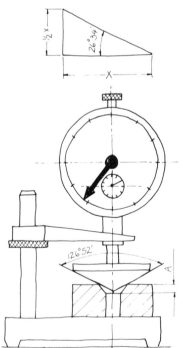

Countersunk dia 4A

Diameter at the surface is generally one of the controlled variables in machining a countersunk hole, the other being included angle, of course. Yet this diameter is not an easy one to measure, and there is no standard instrument, such as a vernier caliper or micrometer, to do the inspection job. But almost every shop has a dial indicator and a vertical stand on which it can be mounted. With proper "tooling," this combination does the job admirably and economically.

What's required is a cone point with an included angle of exactly 126° 52'. This conical tip, which should preferably be hardened and ground, replaces the standard ball-nosed tip on the dial indicator.

Working principle of the device for measuring countersink widths is simply that at 126° 52', the axial length of the cone is precisely one-fourth of its base diameter (tangent of half the point angle is 2.000).

To measure a countersunk diameter, therefore, just zero the indicator with the cone just resting on the flat surface of the work, then move the point so it fully enters the countersink, note the vertical travel on the dial, and multiply by four for the value of the diameter.

        C R NANDA, *Bombay, India*

10.17 Gage quickly measures keyway depths

The sketch shows a keyway depth gage that was made in about an hour and saved a great deal of inspection time at the machine as well as avoiding the need to build a complicated setup for checking production parts. All parts were made of cold-rolled steel, and none of the dimensions are critical; the gage is simply made to suit the particular needs of the individual application.

To set the tool, a master workpiece (gage-piece) is made, and the gage is set by rocking or sliding it back and forth to find the low point in the key seat; then the indicator dial is set to zero.

In using this type of gage, it's important to take into consideration the tolerances on the reference surface. A ground surface to fit the ID of a ball-bearing would be ideal. But if the key seat has to be checked before grinding, it's necessary to check the diameter of the reference first and add or subtract half of the deviation from nominal diameter to the indicator reading.

        WILLIAM WRISLEY, *Liverpool, NY*

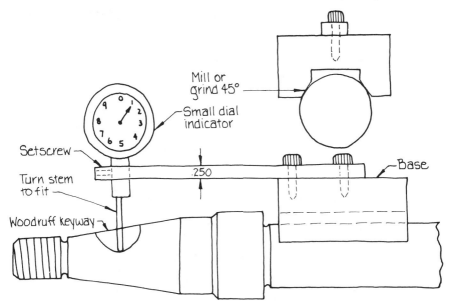

10.18 Horseshoe magnet is indicator mount

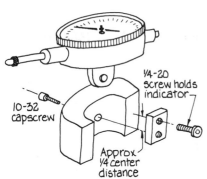

This magnetic indicator mount is so handy that most of my co-workers now have similar holders in daily use. A small piece of 1/2-in.-thick brass, drilled and tapped as shown and screwed to a common horseshoe magnet, makes a very useful and compact mounting for my Starrett long-range indicators—and it fits most other makes of indicators as well.

The 1/4-20 screw for mounting the indicator is a flat-head socket-type screw with the head machined down to about 1/16-in. thickness.

The indicator can be mounted facing either left or right, and, so mounted, it can be used on most lathe carriage ways or the saddle ways of a small mill. It's especially handy on Hardinge lathes.

JOHN URBAS, *Cannon Tool Co, Canonsburg, Pa*

10.19 How to check tapers on the level

In our quality control department we check external conical tapers in a V-block with assistance from a sine plate, gage blocks, and a dial indicator. Laid in the V-block, the cone makes two-line contact for the entire length of the block.

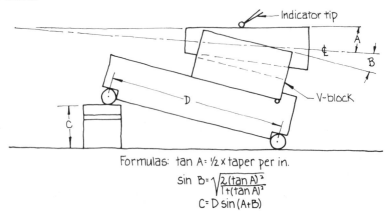

Formulas: $\tan A = \frac{1}{2} \times$ taper per in.

$\sin B = \sqrt{\frac{2(\tan A)^2}{1+(\tan A)^2}}$

$C = D \sin(A+B)$

To obtain a setting, we find the sine of the sum of angles A and B, angle A being half the taper angle and B being the angle from the cone's center line to a line parallel with the V-block.

The gage blocks, of course, are selected to bring the top surface of the cone precisely horizontal. The dial indicator should then read zero along the length of the cone if the taper is correct.

ERNEST GOULET, Middletown, Conn

10.20 How to measure three-flute tools

Ask any toolcrib attendant which is the most unpopular type of end-mill in stock and the immediate answer will be: A reground three-flute end-mill. Nobody wants to use one because its diameter cannot be measured with a standard micrometer, and few machine shops possess the special mike needed to measure three-fluted tools. But there is an easy way to find the size accurately:

Referring to the accompanying sketch, first mike the shank of the tool to determine B. Then roll the tool on a surface plate, using an indicator and gage blocks or a height gage over the cutting edges to determine A. Check all three edges to make sure the tool is concentric. Then calculate the actual diameter of the end-mill by using the following formula:

Dia = B − 2 (B − A)

The precise size can now be engraved on the tool, or a tag or tape attached giving its diameter.

Now when a machinist comes to the crib asking for an end-mill a few thousandths under nominal size, or an oddball diameter for some special slotting job, the attendant can hand over one of these premarked tools.

JOSEPH D JUHASZ, Michigan City, Ind

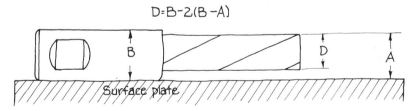

10.21 Indicator adapters clamp on end-mill

There are cases in machining when it's desirable to indicate or reindicate a part after taking a cut but it's desirable to leave the end-mill in the spindle to save

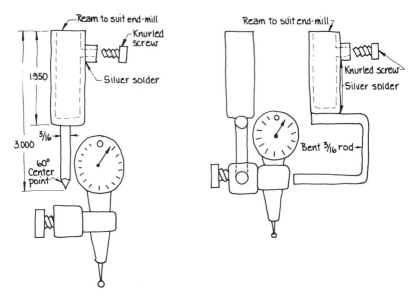

time and not disturb its precise position. These two indicator adapters allow just that.

Both adapters provide a snug fit over the end mill and clamp to it by means of a knurled screw in a threaded boss silver-soldered on. One provides a concentric axial mounting for the indicator; the other provides a mounting perpendicular to the spindle centerline. And it's a quick job with either one to slip the adapter onto the end-mill, check whatever needs checking, and then slip it off to resume machining.

RON STANWICK, *Englishtown, NJ*

10.22 Indicator stand checks squareness

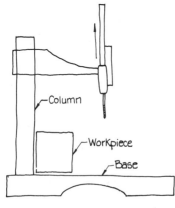

I find that a quick, practical, and accurate way to check parts for squareness on die sections or production runs is to use the base of a dial-indicator stand as a square. It's always handy at the bench, and the column is square to the baseplate. Just slide the part on the base and against the column, and sight this for squareness, as shown in the sketch.

GERARD CATALDO, *Revere, Mass*

10.23 Layout by calculator

When working with a rotary table, I have found that a pocket calculator is a quick and precise aid in indexing. For example, to put 32 holes in a bolt-circle at 11° 15', enter 11.25 into calculator memory. Then, pressing memory and plus-sign buttons continuously, most calculators progress to 22.5, 33.75, 45, etc., until all 32 holes are entered and easily read out by the operator. The operator will quickly adjust to the use or decimal degrees instead of degrees and minutes, and human error is virtually eliminated.

Not all calculators operate in precisely the same way; mine is a Radio Shack Model EC 498.

JOHN J DONOVAN, *Chelmsford, Mass*

10.24 Length gage lightens the inspection load

Measuring the location of the end of a knurled section on a steel shaft on our production line had to be done by an inspector with optical equipment, because of the tolerances involved, until we made the hand-held cylindrical gage shown in the drawing, which allows the checking to be done at the line.

The gage has a spring-loaded, shouldered pin, which retracts in the bore upon insertion of the shaft to be checked until the slightly enlarged OD of the knurl stops further entry. The operator then visually determines if the end of the retracted pin is in the "accept" area between the two lines scribed across the edges of the window milled in the tube.

ROBERT J MCMASTER, *Winchester, Mass*

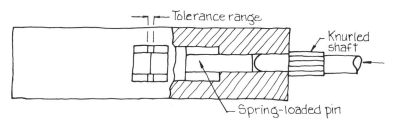

10.25 Lightweight gage for large measurements

We are often required to machine large step diameters—sometimes over 2 meters (79 in.)—on thick, solid plates to rather close tolerances. Measuring these ODs can become a problem if no suitable outside micrometer is available. And fabricated special gages often prove to be too heavy to give the machinist a proper "feel." Our problem was solved by making an adjustable gage from a length of polished aluminum pipe, as sketched.

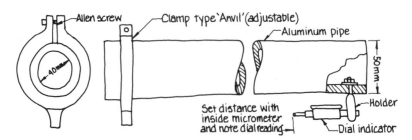

A dial indicator is mounted at one end, with an adjustable clamp-on "anvil" at the other. The anvil's ID and faces are machined square, and it is split on one side to facilitate sliding it along the pipe and clamping it securely.

In use, the gage is set with the anvil positioned at a slightly smaller dimension than the required diameter so that the dial indicator will be displaced by some small increment in measuring the work. A straight line scribed on the pipe ensures alignment of anvil and indicator. Calibration for any particular job is done with an inside ("stick") micrometer set to the required dimension; then the indicator is adjusted to some convenient reading, which is noted for use by the machinist.

By reversing the dial indicator, it's also possible to use the gage for ID measurements.

For the utmost in precision, it's important to remember that the thermal expansion of aluminum is approximately twice that of steel.

M N SARWAR, *Bombay, India*

10.26 Low-profile V-blocks

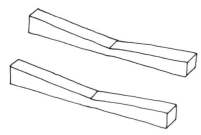

Because the table of our optical comparator only goes a few inches below the center of the working field, only a small, round piece could be placed in standard V-blocks. To get maximum accommodation of diameters, we made up a pair of low V-blocks as shown in the sketch.

MAURICE CONKLIN, *Mississauga, Ont, Canada*

10.27 Measuring machine provides layout assist

Laying out templets, jigs, tools, and fixtures is always time consuming. If your shop has a coordinate-measuring machine, it can speed the process and increase the accuracy at the same time.

For straight lines, take a length of drill rod with the same diameter as the shank of a standard probe; grind a sharp, concentric cone on one end; harden the tip; put it in the spindle; and use as a scriber. Just move the machine, using its digital readout, to the desired dimension, clamp the axis, and you can scribe the needed line by sliding the other axis.

Layout for hole centers can be even easier. Just buy an automatic centerpunch and adapt it to the spindle by either grinding it down or enlarging it with a sleeve—being certain in either case to keep it concentric with the point. Now position the spindle according to the coordinates, clamp both axes, and press down on the spindle to actuate the center-punch. It's a great aid for accurate layouts, especially bolt-hole circles and other jobs that are tedious to lay out.

JOSEPH D JUHASZ, *Willow Grove, Pa*

10.28 'Memory' device aids awkward caliper readings

Measuring a part on a machine with a caliper and then finding it impossible to read the dial or vernier because of the nature of the work or the setup is a frequent frustration that led me to making the device illustrated here. When the situation demands that the caliper setting be changed to remove it from the work, the little clamp on the caliper's depth rod, when set, allows the caliper to be returned to its measurement setting.

Both the clamp and the adapter are made of stainless steel, and are hardened and ground. This auxiliary slide takes about 1/4 in. of space on the depth rod, and thus it cannot be used when the dimension being checked is less than this. However, if you frequently run into the kind of problem the slider is designed for and typical dimensions are this small, it would also be possible to make a new depth rod a bit longer to allow the caliper to be closed all the way. But this would not permit the caliper to be used for conventional depth measurements.

To use the device, close the caliper on the work and lock it in the conventional manner. Then bring the auxiliary slide up against the back end of the caliper and clamp it with the thumbscrew. Finally, open the caliper for removal, reset it against the auxiliary slide, and make your reading.

ERNEST J GOULET, *Middletown, Conn*

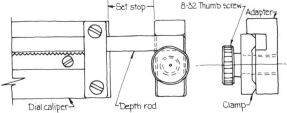

10.29 Microcomputer program for bolt circles

As the owner of a microcomputer, an Apple II, I find AM's articles on CAD/CAM and computer applications of particular interest. And I have developed some programs that should be useful to many small shops using a microcomputer; one such program is listed below.

```
05  REM POINTS
10  PRINT "This program calculates the X and Y coordinates of a bolt
    circle with any number of equally spaced holes. You will be asked to
    type in answers describing the bolt circle to the computer."
20  PRINT
30  PRINT "After typing in the appropriate answer, press the RETURN
    key to continue."
40  PRINT
50  INPUT "Type in the radius of the bolt circle    RADIUS = ";R
60  PRINT
70  INPUT "Type in the number of holes in the bolt circle    NUMBER
    = ";N
80  PRINT
90  INPUT "Type in any X- or Y-axis offsets to the center of the bolt
    circle. In most cases, there are no offsets to the center position, and
    the value is entered as 0,0    OFFSETS = ";X0,Y0
100 PRINT
110 INPUT "Type in the angular location (in degrees) of the first hole
    from the positive X-axis (the hole on the right side of the bolt circle,
    just above the centerline). Type in the angle that would locate this
    hole    ANGLE = ";C
120 PRINT
130 RC = .017453292
140 K = 360/N
150 FOR I = 1 TO N
160     Z = (I − 1) * K + C
170     J = Z * RC
180     X = X0 + R * COS (J)
190     Y = Y0 + R * SIN (J)
200     P = 100000
210     X = INT (X * P + .5)/P
220     Y = INT (Y * P + .5)/P
230     PRINT    TAB( 3);"Hole No."; TAB( 25);"X Coord.";
        TAB( 55);"Y Coord."
240     PRINT    TAB( 6);I; TAB( 25);X; TAB( 55);Y
250     PRINT
260 NEXT I
270 END
```

Layout, Inspection, and Measurement

The purpose of this particular program is to give shop personnel the capability of calculating bolt-circle patterns on a microcomputer. Written in BASIC, the program is user-friendly in that it prompts the user, indicating just what data is required. No programming experience is required for its use, and it will run on any computer that "speaks" BASIC. *[Note: One of AM's editors confirmed this by checking it out on a Radio Shack TRS-80 Color Computer that he shares with his 14-year-old son—Ed]*

Output of the program is a neatly tabulated list of X and Y coordinates for the required bolt circle.

JAMES W PERKINS, *Seminole, Fla*

10.30 Micrometer fixture

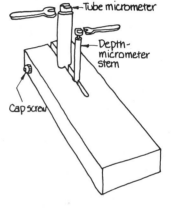

After trying to hold the stems of depth micrometers and tube "mikes" in V-blocks and various other devices to make adjustments, I came up with this simple and easily made fixture that does the job admirably. It's made of a small aluminum block and a socket-head capscrew. The drawing is completely self-explanatory. It can be made in less than an hour.

HANK BAGBY, *Richmond, Va*

10.31 Miking with magnets

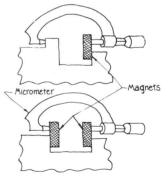

A standard OD micrometer can often be used to measure widths and other configurations that usually require some other type of measuring instrument. Just keep a few magnets handy, and apply them as shown in the sketches. My magnets were purchased in a stationery store; they originally had spring clips attached for holding papers.

JOHN R MAKI, *Beverly, Mass*

10.32 Non-square 'squares' are useful, too

Very handy for die work, molds, models, and other jobs involving small angles is a set of extra blades I made to fit inter-

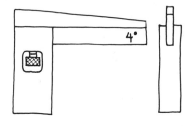

changeably on my 4-in. adjustable square. They aren't square, however, but are ground to provide angles of 1°, 2°, 3°, 4°, 5°, and 7°, which are the ones I work with most often.

They can be used with a surface plate for layout and inspection. But they are even more useful for setting up thin workpieces in a milling-machine vise (when standard angle blocks are just too thick for the work). Blade thickness is about 0.065 in.

RON STANWICK, *Somerville, NJ*

10.33 Parallelogram fixture centers workpieces

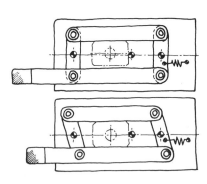

It is sometimes necessary to center workpieces—for drilling or piercing, for example—to closer tolerance than that of the part's width. To do this with a nesting type locator, you've got to tighten up on the width tolerance, which would otherwise be unnecessary.

The parallelogram cage shown on the drawing will do the job handily for many workpiece shapes—square, rectangular, hexagonal, round, etc—and quite a range of sizes. The principle can be applied to press tooling, drill jigs, milling fixtures, even inspection devices and gages. And it's simple to make.

L PAKKRISAMY, *Madras, India*

10.34 'Parallels' adjust for uneven surfaces

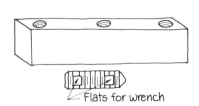

I have made up a pair of parallels that come in very handy when it's necessary to machine a casting or any other job with an uneven surface that has to rest on parallels. These parallels have adjustable support screws that permit the casting to be leveled for clamping to the table and machining.

The adjustable parallels are made of 3/4 x 1 1/2 x 10 in. tool steel with 3/8-16 holes tapped in the center and 1 in. from each end. The parallels are hardened and ground. The adjusting screws are made 1 3/8 in. long from 3/8-16 threaded bar stock. One end is faced off flat; the other is turned to a 90° included angle with the point rounded off. In addition, flats are ground near each end of the screws for an adjusting wrench. The screws can be used with either end up, depending on the surface to be contacted.

Obviously, the parallels and screws can be made up to any length required for the typical jobs in any given shop.

JOHN URAM, *Cohoes, NY*

10.35 A plastic center-finder

In layout work it's often necessary not only to locate the exact center of existing holes, but also to strike off dimensions from those nonexistent points. A small piece of transparent plastic, as shown in the sketch, can be very useful for both purposes. A small hole is first drilled in the plastic, and a circle matching the diameter of the hole in the workpiece is scribed about this point. Then it's a simple matter to lay the plastic over the hole (scribed side down for best accuracy), lining it up carefully, and then laying out the necessary dimension from the hole in the center with a pair of dividers. The plastic center-finder can be used over and over again, simply scribing new diameters as they're needed.

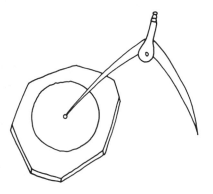

RICHARD GRIFFIN, *Adrian, Mich*

10.36 Proximity probe makes good 'centricater'

Some nuclear tubes required internal boring of two diameters. The requirement was that these bores had to be concentric within 0.01 mm (0.0004 in.). Machining was not much of a problem; but to prove the concentricity to the customer's inspector became a problem because the bores were only 30 mm

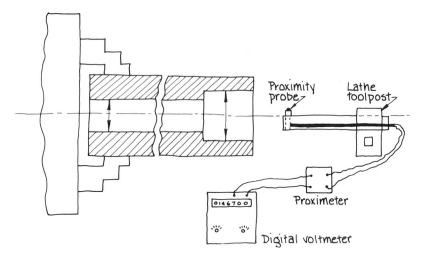

(1.181 in.) and 40 mm (1.575 in.) in diameter and no dial indicator would go in.

Rather than make a special mandrel or stepped plug gage, we checked with our research and development department and found that they had some proximity probes that would fit within the bores. We borrowed their smallest unit, fitted it to a boring bar, and calibrated it for the material of the tube. It provided a linear output on a voltmeter when the gap between the probe tip and the metal was between 0.5 and 1.6 mm (0.020 and 0.063 in.). We used a digital voltmeter to sharpen the precision and got the job proved.

C R NANDA, *Bombay, India*

10.37 Radius of segment on a profile projector

When you have only a segment of a circle and you have to find the radius, this simple formula does the job:

$$R = \frac{A^2 + B^2}{2B}$$

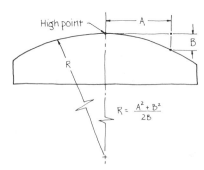

'R' is the radius you're trying to find, and 'A' and 'B' are the dimensions indicated in the drawing. To find A and B, the high point of the segment must be centered on a contour projector; then A can be taken as any convenient dimension you want, and B is the dimension from the tangent line at A to the circular arc. We use a 10:1 magnification with digital readout. Several readings will increase the accuracy when the mean (average) is taken.

To prove your technique and gain some confidence, try the method on a segment with a radius that you know.

ERNEST J GOULET, *Middletown, Conn*

10.38 Recondition that worn scale

One of the most-used tools in the machine shop is the 6-in. steel scale. Because it gets used for many purposes other than measuring, the end corners may become rounded. When this happens, I renew my flexible scale by clamping it edge-up between two parallels and grinding off 0.030 in. or so until the corners are once again square. By wet-grinding it, you can avoid any discoloration. And it's as good as new.

HARVEY SCHULZ, *Hillside, Ill*

10.39 Reversible protractor for the toolroom

This reversible protractor seems almost indispensible for a wide variety of uses in setting up work on surface grinders and milling machines. The precise angle to be used is set with the aid of a conventional protractor, and then this one can be used against the back rail on a magnetic chuck, for example, to position the workpiece at the desired angle.

Dimensions are not all critical, but mine is made of a 4-in. length of oil-hardening steel 7/16 in. thick and 1 3/8 in. wide. The slot is milled 5/8 in. deep, but less than 0.100 in. wide to allow for final grinding all over to make all surfaces flat and square. The blade, also hardened and ground, is 0.100 in. x 1/2 in. x 3 3/4 in., drilled at one end for the body diameter of the 8-32 Allen-head capscrew that holds the instrument together. The protractor base is drilled and tapped on one side for this, and the other side is counterbored to accommodate

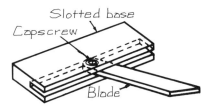

the head. When the protractor is being set, the screw thus clamps down on the blade.

RUDOLF K SCHMITT, *Whitestone, NY*

10.40 Scriber follows edge, straight or curved

The illustrations show a simply made scribing tool that is extremely useful for layout of such details as bolt-hole patterns that must be a specific distance in from the periphery of such curved workpieces as a circular flange. At the same time, the scriber will also follow a straight-edged part.

The instrument can be made from virtually any scrap stock lying around the shop, and the precise shape of the body is not particularly important. Also, dimensions can be virtually anything that suits the general size of workpieces typically machined in your plant.

ROBERT S MOSS, *Clifton Park, NY*

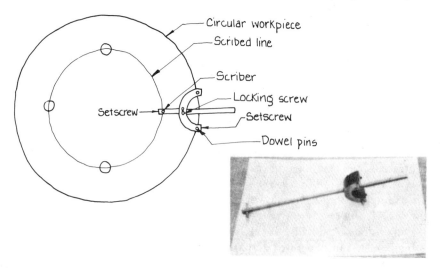

10.41 Scriber is 'ambidextrous'

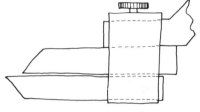

Two 1/4-in. toolbits are first ground to suit the user's personal preferences (I use a full radius on the ends) and are then set up face-to-face in a height gage, as illustrated. This allows measurement or scribing either "up" or "down" with no adjustment other than loosening the clamp screw and sliding one bit out slightly farther than the other for clearance. It's especially handy, I find, when used on an "Instant Zero" type of digital or dial height gage.

DAVID R CARLSON, *Manchester, NH*

10.42 See-through templets speed layout

Transparent templets made of Plexiglas or other fairly rigid but optically clear plastics can greatly simplify the layout of workpieces that must be a face-to-face match with existing components. You just lay the plastic on the surface to be matched and "trace" it. Transfer points and scribers can be used just as you would if making the templet from sheet metal.

LEE C WILKERSON, *Keyesport, Ill*

10.43 Setting drill-point depth

In programming numerically controlled machines we have to know everything about the drill point to attain the proper depth when we tell the operator to preset the drill point against the work or a predetermined height above it.

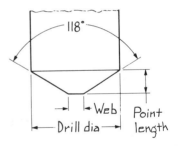

Subtracting web thickness from drill diameter and multiplying this difference by 0.3 gives the point length with sufficient accuracy for a standard 118° drill point. The 0.3 is approximately one-half the tangent of 31° (tan 31° = 0.60086).

ERNEST J GOULET, *Middletown, Conn*

10.44 Shimming V-blocks to level stepped shafts

When a stepped shaft has to be supported on V-blocks for inspection, there's a very simple formula to determine the height of the shim ("C" in the drawing) to be placed under the V-block at the smaller diameter to maintain the shaft in a horizontal position.

The value of C is A − B, each of which is the square root of 2 multiplied by respective radius. Hence:

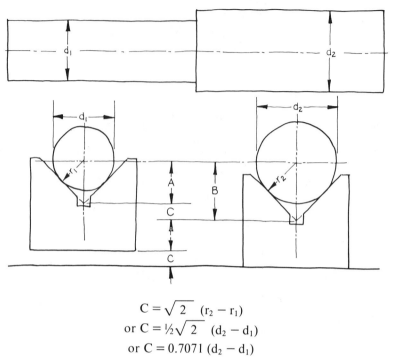

$$C = \sqrt{2}\ (r_2 - r_1)$$
$$\text{or } C = \tfrac{1}{2}\sqrt{2}\ (d_2 - d_1)$$
$$\text{or } C = 0.7071\ (d_2 - d_1)$$

It's that simple!!!

M F SPOTTS, *Evanston, Ill*

10.45 'Siamese-twin' angle plate is versatile

The need often arises to hold small workpieces for milling, drilling, grinding, and even such hand work as deburring. The illustrated fixture combines the functions of an angle plate and a V-block into a single unit resembling a pair of angle plates joined base to base at 90°. Made of s7 steel (hardened and ground), it holds either round or square pieces accurately and with considerable versatility. And it saves considerable shop time because of the quick setups it makes possible.

The numerous holes, some tapped for 5/16-18 bolts and others simply clearance holes for the same bolt size, provide many clamping positions on every surface, as well as making the tool lighter and relieving internal stresses. Further, they provide positions for shoulder bolts, which make excellent work-positioning stops.

The fixture can be clamped in a machine vise or can be bolted directly to the table. And it can be used at the edge of the table, allowing long workpieces to hang over the side for machining operations on the ends.

Layout, Inspection, and Measurement

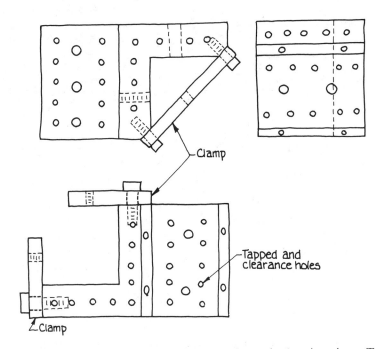

Various styles of clamps can be made to suit particular situations. Two of these are illustrated: one bolted to the 45° chamfers with a central tapped hole for a screw for clamping round stock, and the other bolted to an edge, also with provision for a clamping screw.

One useful application is when a piece has to be ground square on all six sides. After two sides have been ground parallel, the fixture holds it square for grinding the third and fourth sides. Then the usually difficult fifth side can be ground square with the part again clamped within the "L."

JAMES ALBINGER, *Saukville, Wis*

10.46 Simple gage measures keyway offset

Here is a simply made gage that can be used to check and measure the offset of milled keyways in shafts—without taking the shaft to a separate inspection department, and often without even removing it from the milling machine fixture. As such, it's quite useful both for operator inspection of parts being produced and for initial setup of the milling machine.

Operating principle is simple. The inverted V-block sits on the periphery of the shaft, and alignment is not influenced by variations in diameter. The eccentric pin in the bottom of the indicator piece, when inserted in the keyway, then "cranks" the pointer one way or the other to indicate alignment or lack of

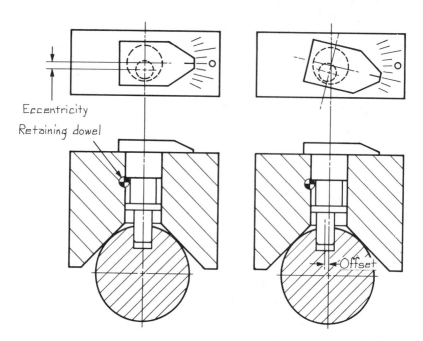

it. The fact that the pointer is longer than the eccentricity of the driving pin magnifies the offset for easier visibility, and calibration of the pointer is not difficult.

E AMALORE, *Madras, India*

10.47 Sine bar uses standard dowel pins

You can make yourself a low-cost sine bar, as shown in the drawing, by using standard 1/2-in. dowel pins. The bar itself can be made of either tool steel

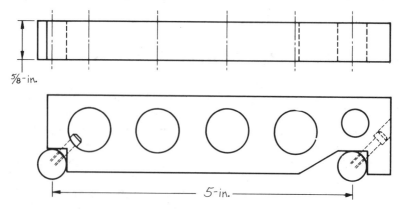

(which should be hardened) or cold-rolled steel (which should be case-hardened). Surfaces, of course, should be ground. Standard dowel pins have a soft core, so they can be tapped if a small area of surface is first removed. Obviously, the sine bar will be only as accurate as you make it.

EDMUND JEZUIT, *Chicago, Ill*

10.48 Special fixture calibrates machine dials

We build special machinery and occasionally need a few special calibrated dials. We developed a special fixture that permits them to be produced quickly.

The basic component is a gear with a number of teeth equal to or a multiple of the number of divisions to be scribed on the dial. A pin is pressed into a baseplate to permit the gear and the dial to rotate on it (these being keyed together by a small pin). This assembly is then set up on an arbor press (or even a drillpress).

The cutter assembly consists of a shank turned to fit the end of the arbor press ram (or drillpress chuck) and a crossbar welded to it that carries the scribing-tool bit at one end and a tapered (to fit between the gear teeth) plate adjustably at the other end.

Length of each line to be scribed is controlled by the length of pins set vertically between the gear teeth and secured by a hose clamp or even a heavy rubber band.

Operation is fairly obvious.

CLINT MCLAUGHLIN, *Jamaica, NY*

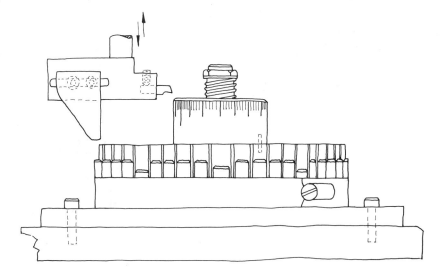

10.49 Special gage is simple to make and use

We recently had a job order for many pieces of three-lugged work that required measuring of a close-tolerance root diameter. The arrangement shown in the illustrations worked well and also allowed measuring into the sharp inside corners of the work.

A Starrett 1-in. micrometer head was used in a ring-shaped frame opposite two adjustable anvils, which were made by cutting a drill blank in half and grinding the ends square.

A setting gage, also shown in the sketch, is required for positioning the adjustable anvils for each diameter to be measured. The micrometer head that we used was calibrated normally for a single opposed anvil, rather than for twin anvils at 120° spacing; thus readings taken at some distance from the

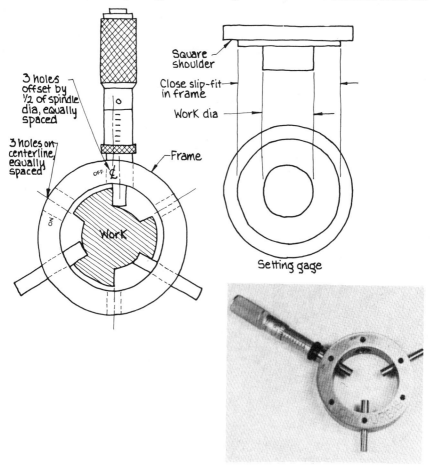

setting established by the setting gage are exaggerated. A short length of bar was turned to the large and small tolerance limits in order to establish comparison readings.

If I were to make another one, rather than using setscrews to hold the micrometer barrel, I would split the frame with a saw cut parallel to the plane of the ring and through both micrometer-mounting bores so that a single capscrew could be used for clamping the micrometer in either of its positions.

JOHN URBAS, *Cannon Tool Co, Canonsburg, Pa*

10.50 Special snap-gage for obstructions

Maintaining close tolerances on the diameter of an obstructed face groove like the one indicated as 26.525/26.527 in. on the sketched part can be quite troublesome. You can't get at it with a standard micrometer or vernier caliper; there's no other dimension it could be referenced to, such as a bore or an OD; and a special recess mike would be too expensive.

We solved just this problem by making the special snap-gage shown at the bottom of the sketch. The frame is made of steel plate, flame-cut, cleaned, and straightened. The surfaces on which the pin-shoulders seat are milled level, and two holes are drilled at a center distance calculated to give the correct dimension over the pins when they are fitted.

To avoid the need for jig-borer accuracy, one of the pins is made eccentric, as shown, so that it can be adjusted to the precise dimension desired with the aid of an outside micrometer.

P MURALIDHARAN, *Madras, India*

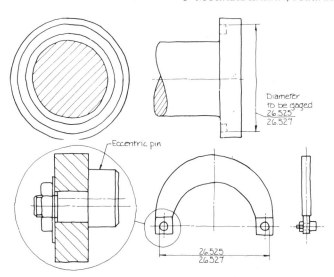

10.51 Split nut gages threads

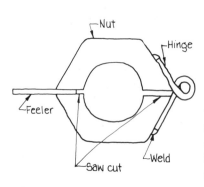

For turning large-diameter threads, a simple gage made in the shop can help the lathe operator. Simply weld a hinge on one side of a suitable nut and then saw the nut in half, as sketched. The nut is merely closed over the work, with a feeler gage used in the open end to allow for the kerf thickness and to tell the operator when the thread is approaching correct size. This allows the operator to check work in progress without backing off the tailstock for checking with the finish gage until absolutely necessary.

MARTIN J MACKEY, *Parma, Ohio*

10.52 Spotting punch is centerline scriber

The familiar spotting punch is normally used to locate and mark the center of a hole in line with the clearance hole it passes through. This is the general method of locating a hole to be tapped in a matching component. But what if the clearance hole is a slot that has been cast or milled to a width that is not standard for your spotting punch and you have to tap-drill a series of holes centered on the nonstandard slot?

Use a spotting punch about 1/16 in. larger than the slot and grind flats about 1/32 in. deep—but of equal depth—on opposite sides so that it's a free fit in the slot. Then twist the flatted punch in the slot so that opposing corners of the flats contact the inner walls of the slot, and pull the punch along the length of the slot so that its point scribes a center line. The opposing corners will keep the point centered. You can also use the punch in its twisted position for spot punching marks for subsequent centerdrilling and drilling.

BEN SCHNEIDER, *W Orange, NJ*

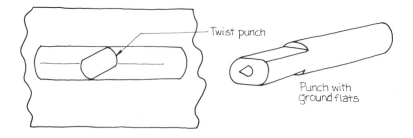

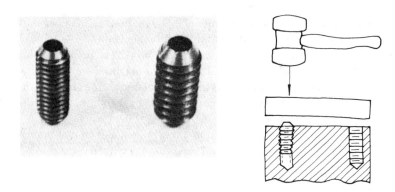

10.53 Standard setscrews as transfer punches

There are many times in the machine shop when the need arises to spot a hole or transfer a location from a tapped hole in a mating part. Lacking a standard spotting screw of the proper thread, a perfectly adequate job can be done by grinding a chamfer on the socket end of a setscrew—such as the 10-32 x 1/2-in. and 1/4-20 x 1/2-in. setscrews in the photo—and then screwing that into the existing threaded hole until only about 0.015–0.020 in. projects from the surface, and finally locating that part on the one to be drilled and tapping it with a soft-faced hammer. This leaves a mark that's easy to locate on, and there's no trouble removing the screws.

Ron Stanwick, *Englishtown, NJ*

10.54 Step-block, scriber give built-in sizes

The combination of this special step-block and scriber, used on a surface plate, permits accurate scribing of lines at precise 1/16-in. increments.

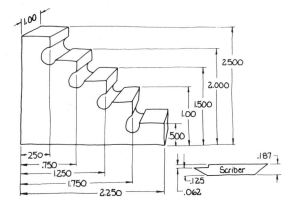

The step-block provides increments of 0.500 in. in one direction and 0.250 in. in the other, yielding exact 1/4- and 1/2-in. dimensions up to 2.500 in. The scriber, with its special configuration, can then be used in any of four orientations to add 0.000, 0.062, 0.125, or 0.187 in. to the height of the selected step on the block. And if the layout requires incremental dimensions in steps of other than 0.062-in. height, a feeler gage can be used beneath the scriber for measurements to within 0.001 in.

The scriber is made of a 3/16-in. square toolbit. And, of course, each step in both the block and the scriber should be ground accurately.

ROBERT F BOWEN, *Burlington, Mass*

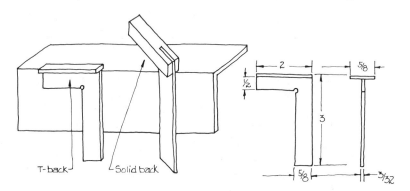

10.55 T-back square is like having a third hand

Maybe someone sells a square like the one in the drawing, but after a brief search I gave up and decided to make one for myself. I machined mine out of a solid piece of tool steel, and then hardened and ground it. The dimensions, of course, are completely optional.

What makes this square so practical is that it lies flat on the surface, with its T-back lining up against the edge of the workpiece and supporting itself instead of dangling like an ordinary solid-back square. Everybody in the shop likes this handy tool the first time it's used to scribe a line.

ANTON MESSERKLINGER, *Everett, Wash*

10.56 Tubular screw jack

When running certain tests and when aligning test equipment, it is often necessary to have a surface readjusted to a new height. We had been using blocks for this purpose, but often they had to be machined or we had to use shims in addition to existing blocks. It was messy and time consuming, and it required a collection of blocks and shims.

To solve the problem, I designed an easily made adjustable tubular screw jack. As shown in the drawing, it's made with a standard hex-head bolt, which is drilled axially to clear a socket-head capscrew. The bolt head is also faced off square. A length of hex stock is drilled axially and tapped to accept the hex bolt.

Screw sizes, of course, depend on the specific needs of the job at hand. In our case, I used a No. 10 capscrew and a 3/8-24 hex bolt.

The screw is simply inserted through the adjustable surface and the tubular jack, and threaded into the base plate. The height can then be positioned with considerable accuracy. The tubular screw jacks can be adjusted and checked with a micrometer.

ART DRUMMOND, *Walworth, NY*

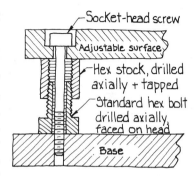

10.57 Two edge finders are easier to make

This tool is useful for quickly locating the center of a collet or chuck in relation to the edge of a workpiece and similar setup jobs.

I made mine to the dimensions shown, hardening it between the machining and grinding operations, and finally cutting it apart to produce two identical tools. The advantage of making the tools as a pair is that the ends can be turned and ground on centers.

FRANCIS J GRADY, *Reading, Pa*

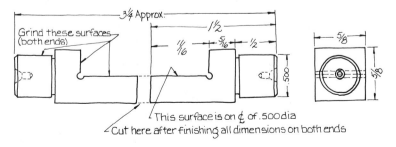

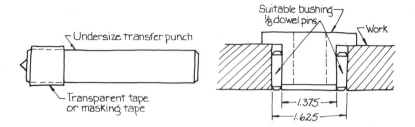

10.58 Two ideas for transfer of center points

The sketches show two ideas that come in handy when you have to transfer a center location from one part to a mating part—and you don't have just the right size items. First, an odd-sized transfer punch can be made up simply by wrapping an undersized transfer punch with a few layers of transparent tape (such as "Scotch" tape) or masking tape until the diameter is suitable. And, second, if you have to transfer centers from a bushing that's too small on its OD, you can "adjust" it with short lengths of drill rod or dowel pins. Three or four equally-spaced pins will do the job.

THOMAS TOMALAVAGE, *Roselle Park, NJ*

10.59 Versatile V-block sits, stands on head

I often find uses for the versatile V-block shown in the drawing. It's easy to use and can quickly be set up in a variety of ways, and it's especially handy when the workpiece cannot be held securely in a standard machine vise. Thus the temptation is avoided of making a haphazard setup.

The drawings show two of the ways it can be set up, one for holding round stock vertically, the other horizontal. And the V-block itself can be held in the

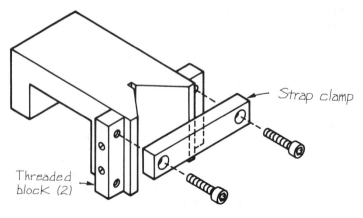

Layout, Inspection, and Measurement

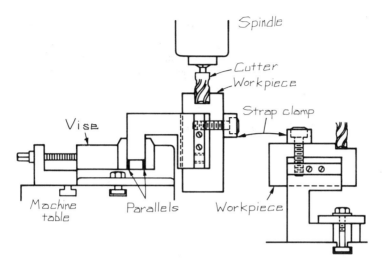

jaws of a standard machine vise or can be fixed directly on the machine's table with strap clamps.

EUGENE CLINTON, *Warren, Pa*

10.60 Diameter of odd-numbered bolt-circle?

It's a cinch to measure the circle diameter of an even-numbered bolt circle; just insert pins (slip fit) in two opposite holes, measure across the outside, and subtract one pin diameter. But when there's an odd number of holes, it isn't that easy—but it doesn't have to be that tough, either, especially with hand calculators now so widely available.

To find the diameter for a bolt circle with an odd number of holes, you can still insert two pins, but this time slip them into holes that are adjacent. Again subtract one pin diameter from the measurement—and then multiply the resulting figure by a suitable constant from the table given with the accompanying drawing.

For example, if A in the drawing works

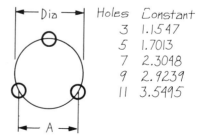

Holes	Constant
3	1.1547
5	1.7013
7	2.3048
9	2.9239
11	3.5495

out to 3.464 in., multiplying by the three-hole constant of 1.1547 gives a diameter of 3.9999 in.—or call it 4.000.

ERNEST J GOULET, *Middletown, Conn*

11 Joining and Assembly

11.01 Bushing plug prevents pressing distortion

When a bronze or Oilite bushing is pressed into a bore, it will often close up slightly in ID and become undersize for the shaft it's supposed to fit. It's then necessary to ream it or remachine it to fit the shaft.

This can be prevented by first inserting a dowel pin of the same nominal size as the ID of the bushing and then pressing the bushing into place. The dowel pin, being 0.0002 in. over its nominal size, will allow the shaft to slide in freely after it has been tapped out.

JOHN URAM, *Cohoes, NY*

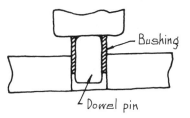

11.02 Calibrate that wrench

An adjustable wrench is a very handy tool, but sometimes the smaller size of an open-end or box wrench is necessary because of tight space around the nut or bolt being wrenched. The adjustable wrench is often applied first, and then the decision is made to use a different wrench type.

But what size is that hex?

A quick measure can be obtained if your wrench has a simple scale to show jaw opening, as illustrated. The scale can be marked on the wrench with a vibratory pencil, electrochemical etch, stamping, etc.

It's no vernier caliper, but it will save you some unnecessary trips back to the toolbox

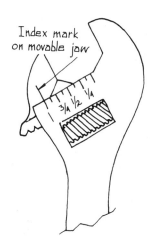

whenever you guess wrong on the size wrench you need.

MIKE HERVEY, *Wilmington, NC*

11.03 Color codes separate parts

Assembly workers can immediately identify small but similar components if they're color-coded. Sieve-dip the parts in dyes of different colors. Often only one color is necessary to solve an assembly problem—and you can always use the ever-available machinist's bluing.

WILLIAM SLAMER, *Menomonee Falls, Wis*

11.04 Driver aligns bushing

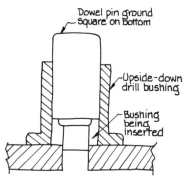

This simple setup is used to square up and insert bushings into plates. It consists of a long drill bushing with an ID larger than the head diameter of the bushing being inserted and a dowel pin ground square on one end. The bushing being inserted is positioned on the plate, and the larger bushing with the dowel is placed over it. The dowel is lowered until it sets on top of the bushing being inserted, which squares up the bushing. The bushing can now be driven or pressed in without the chance of shearing due to misalignment.

HOWARD THURSTON, *IBM Corp, Endicott, NY*

11.05 Friction pin aids assembly

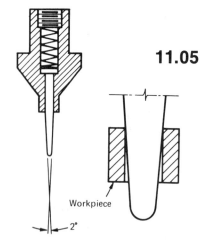

There are many times when it's easier to load a small part on the punch for an assembly operation. The problem is holding the part on the punch before it descends into the assembly. When a part has a center hole I have used split pins, magnetized pins, and greased pins. These work, but the split pins lose their resilience, the magnetic pins get fuzzy, and greased pins are prohibited in some assemblies.

The illustration shows a spring-loaded taper pin that holds a part simply with friction. The taper is so slight that the part is held square with the assembly. Correct spring pressure is important, however, for easy release.

ERNEST J GOULET, *Middletown, Conn*

11.06 A handy hex-driver

About 2 1/2 in. of aluminum barstock—I used 1-in. dia 6061 alloy—makes a very useful tool for driving sheetmetal screws with 1/4-in. and 5/16-in. hex heads, which are commonly found on appliances, electrical equipment, and air-handling equipment. Ends of the bar are drilled and tapped for 3/8-16 and 5/16-18 socket-head capscrews, which are hardened and thus provide long wear—and are easily replaced when necessary. The locknuts keep the capscrews from turning, and the knurl provides a good grip.

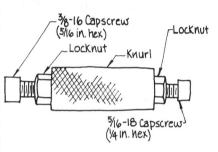

L B SCOTT, *Knoxville, Tenn*

11.07 Handy offset drivers for screws and nuts

There are times in assembly work when you need an offset screwdriver—and none are available. Flats can be ground on the short end of Allen wrenches (suitably sized, of course) to produce makeshift offset drivers. Make two, with the flats 90° from each other, so that the screw can be tightened in quarter-turn steps.

Similarly, an offset socket-driver can be made from a correctly sized socket-head capscrew (as long as possible) by heating it just below the head with a torch and bending it 90°.

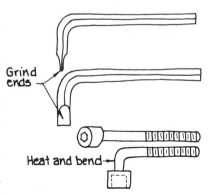

JOHN URAM, *Cohoes, NY*

11.08 A handy pin-presser

The operator was using a hammer to press steel pins into a large plastic frame. My assignment was to eliminate the hammering operation. I purchased one of those spring-loaded Brown & Sharpe automatic center punches. Then I removed the point and made a collet with a small magnet insert to retain the pin while positioning it over the hole in the plastic frame. Downward pressure then releases the striker, which, in turn, drives the pin into the hole.

The solution both reduced assembly cost and eliminated rejects due to bent pins.

JOSEPH P OSCAR, *Stamford, Conn*

11.09 Installing bushings that can't be reached

There are times when it's necessary to press in a pair of flanged bushings or bearings in a yoke—with the flanges on the inside of the yoke. This makes tight work, and it's tricky to keep everything straight and aligned.

I simplified such a task by making up this rod and split collar. The arbor material is drill rod, or a long dowel pin if available, that is a slip fit in the bushings. Length should be enough to press from both sides.

The arbor is first inserted through one leg of the yoke. Then the first bushing, the split collar, and the second bushing are slipped into place. Next, the collar is clamped in the center of the arbor. The bushings are then aligned with the bores and pressed in on an arbor press—first in one direction, then in the other.

The arbor ensures that the bushings are straight and aligned with each other, and after the job is done it's simple enough to loosen the split clamp and slide the arbor out.

JOHN URAM, *Cohoes, NY*

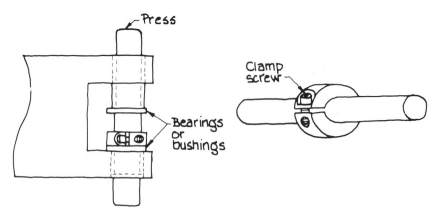

11.10 Instant photos aid assembly

One way to cut costs for printed-circuit-board assembly—and it would be applicable to many other small assembly jobs a well—is to use an instant camera, such as a Polaroid camera. Just take an instant photo of the first board assembled and use that as a guide for additional assemblies.

There are several advantages: The photo is often easier to use by unskilled personnel than a customer drawing with technical component nomenclature. Your records can be "instantly" updated when design changes come along. And the pictures can be in full color.

Sometimes it may be useful to take two photos; one for the assembler and one for the files. And, for larger work, it may be advantageous to enlarge the working photos to 8 x 10-in. size. When we want such enlargements, we generally use a 35-mm camera.

JOHN R MAKI, *Beverly, Mass*

11.11 It's a 'Yankee' hex driver

Having seen the "Handy hex-driver" in a previous Practical Ideas feature, I thought I'd submit an alternative to that idea.

Take a 3/8-16 or 5/16-18 socket-head capscrew (for driving 5/16-in. and 1/4-in. hex-head screws, respectively) and modify the threaded portion to fit a push-type screwdriver—sometimes called a "Yankee" screwdriver. This will require turning the diameter, and grinding the driving flat and retaining notch.

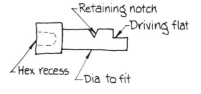

It's likely that many of the technicians who would use such a tool already have a push-type screwdriver in their toolboxes.

STAN SAUNDERS, *Columbus, Ohio*

11.12 Keep that spring in place

When a spring is coiled of small-diameter wire, there is sometimes a tendency for it to slip into the ID of the washer retaining it, causing binding between spring, washer, and screw—and trouble. Simply by unwinding about a quarter turn of the spring,

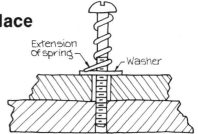

however, this difficulty can be prevented. The end acts to prevent the spring from entering the washer, as illustrated.

BEN SCHNEIDER, *West Orange, NJ*

11.13 Lapping small cylinders

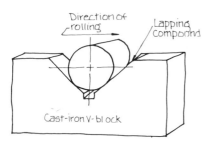

Small, cylindrical, heat-treated parts often have to be sized to provide a precision fit in a bore. A simple, inexpensive, and accurate way to size such a part is to make or purchase a cast-iron V-block, apply a small quantity of diamond dust or lapping compound in the V-groove, and roll the part in the V with your finger. The system does a good job of producing very precise fits.

MARTIN K ADEMOVIC, *Rochester, NY*

11.14 Lathe is convenient press

An unused lathe, and most shops seem to have an old one around somewhere, can be used as a sort of horizontal arbor press for installation or removal of bushings in workpieces. Depending on the specific job at hand, it's usually a simple job to nest the parts loosely in the chuck and use some suitable step stud held in the tailstock to press the bushing in, or through, the part by advancing the tailstock spindle.

JOHN R MAKI, *Danvers, Mass*

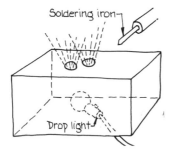

11.15 Light seeks leaks

I've found that the easy way to solder a can or a bucket or the like is to invert it over a light bulb. The light shines through to show where the voids are, and you'll also know when the solder has sealed them.

ALLAN PARADIS, *Kennebunkport, Me*

11.16 Make a handle for small-size welding wire

This easily made handle, similar to a pin vise, provides many advantages for a variety of manual welding and brazing jobs: It allows filler wires to be used

Joining and Assembly

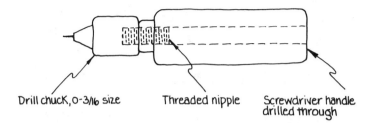

Drill chuck, 0-3/16 size   Threaded nipple   Screwdriver handle drilled through

down to the last 1/2 in. or so; it provides good control of the wire with a gloved hand; it facilitates use of lengths cut from a reel (the tool will straighten it sufficiently); and it protects one's hand from the heat of the job.

Quite inexpensive, the tool is simply made from three parts: an old drill chuck, 0 to 3/16 or 1/4 in. capacity, with a through hole threaded to accept a nipple; a short, threaded tube (nipple); and a file handle or screwdriver handle that has been drilled out axially for the nipple.

NICHOLAS SUDAK, *Saugus, Mass*

11.17 Method for locking screws

To convert any machine screw or bolt to a locking fastener, just rap a section of the threads with a chisel point, as sketched. This distorts the thread and provides a positive locking action on assembly. It's an especially useful technique out on the shop floor.

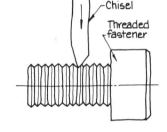

MARTIN K ADEMOVIC, *Scottsville, NY*

11.18 An 'overhead' weld at floor level?

I recently had to close off a 3-in. x 12-in. horizontal opening—pressure tight—in a tank made out of boiler plate. The method of choice was to hold a 4-in. x 13-in. (for overlap) piece of plate over the inside of the hole with a temporary handle for tack-welding, then burn off the handle, and arc-weld the seam along all four edges. The snag, however was that the upper horizontal weld would be an overhead weld—difficult at best—and the opening was only inches above floor level. You'd have to stand on your head to see what you were doing!

The difficulty was avoided by first solidly welding a 1-in. width of the plate (12 in. long to fit in the hole) near one edge of the patch plate. This was done at the bench, downhand. Then the handle was attached and the job was done as

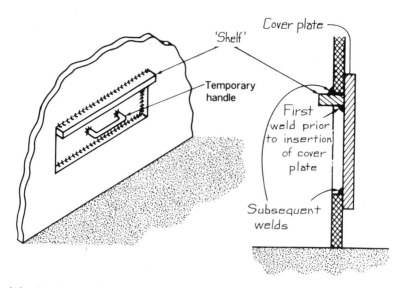

originally planned—except that both the top and bottom seams were now conveniently downhand.

ANDREW VENA, *Philadelphia, Pa*

11.19 Plastic tubing handles tiny nuts

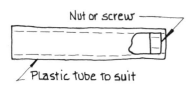

Positioning and assembling small nuts and screws in open spaces is usually difficult enough; in tight places it becomes almost impossible without some kind of mechanical assistance. I have found that plastic tubing—even soda straws—can be almost indispensible in my toolkit. See sketch.

W E TRITZ, *Waukesha, Wis*

11.20 Pliers modified for pulling

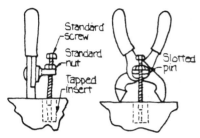

A pair of pliers is a functional, versatile tool. But the pivot pin simply holds the two halves together.

Without impairing the normal usage of the tool, we replaced the pivot pin with a special, slotted, shoulder pin—as shown in the sketch—for removing small, tapped inserts from blind holes. A standard screw and nut make a puller of adjustable length, and closing the plier handles jacks up the insert by the cam-like action of the closing jaws.

ERNEST J GOULET, *Middletown, Conn*

11.21 Plug aids bearing installation

Pressing a ball bearing into a casting or housing can easily result in damage if the job is not done properly and carefully. Especially important are squareness of the bearing with the bore and making sure that pressure is not applied to the inner race. The special plug shown here greatly facilitates doing the job correctly without damage to the bearing.

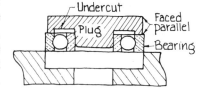

The central plug is turned to a slip fit in the inside diameter of the bearing; its purpose is to ensure squareness. The face of the plug is undercut so that pressure can only be applied on the outer race of the bearing. And the OD of the plug is turned approximately 1/64 in. under that of the bearing so there will be no interference with the housing bore.

Not only does this plug prevent damage to the bearing, it also makes the job of pressing it home in an arbor press much easier.

JOHN URAM, *Cohoes, NY*

11.22 Puller for inaccessible and fragile sleeves

In reconditioning work on one of our products, it is necessary to remove a fragile ceramic sleeve that is accessible from one end only in order to replace a gasket. Most often, the sleeve is tightly frozen in place and is most difficult to remove without damage—and it's expensive to replace. Because it is a recurring problem, this special extracting apparatus was devised and built for the repair department.

Body of the tool is turned from barstock so that it is a slip fit in the ceramic sleeve. Three equally spaced holes are drilled through this lengthwise for

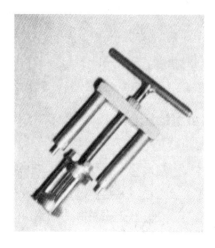 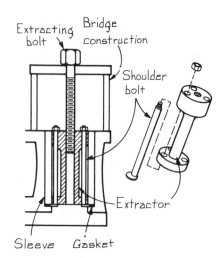

shoulder bolts. The holes must be very close to the body's OD so that the bodies of bolts are near the edge. Heads of these three bolts are then ground down so that they will fit into the space where the gasket is, and they are ground eccentric so they can be rotated into the space.

The tool body is also tapped axially for a pulling bolt that bears on a simple bridge structure, as shown.

ROBERT J PHILLIP, *Oshkosh, Wis*

11.23 Sealant for leaky welds

We have a part that's made of steel tubing welded together and subsequently chrome-plated. Normally we would drill 3/16-in. holes to drain the tube sections that are closed off. The reason for this is that if the weld is not perfect, the plating solution is sucked into the tube through minute leaks by a vacuum created during the welding cycle, and this trapped solution will later leak out slowly and ruin the plated finish. Hence, either drain holes or perfect welds are required.

I recently discovered a very simple technique that eliminates both needs: I painted the welds with Loctite sealant, which is drawn into any minute voids by capillary attraction and which then hardens into a tough plastic in the absence of air. This anaerobic sealant not only prevents entry of the plating solutions, it can even be used to make the welds pressure tight. I have even used it to fabricate hydraulic reservoirs, tanks, and pressure-tight cylinders.

It might be pointed out, however, that this method will not seal large gaps left in a joint as a result of poor welding technique.

DOUGLAS B BETT, *Dana Point, Calif*

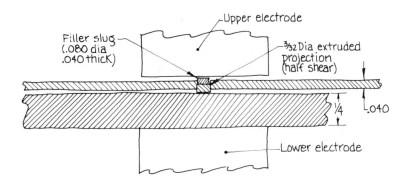

11.24 Slugs enable projection welding aluminum

We recently had a job that required welding 0.040-in. aluminum to 0.250-in. aluminum. The particular application, however, demanded a good surface appearance. At first we tried spot welding, but the surface was indented by the electrode, or burned completely through, making it necessary to scrap both parts.

Projection welding of aluminum is normally not feasible, because the pressure of the electrode flattens out the projection before the heat is applied. However, by using a filler slug when projection welding, as shown in the sketch, we were able to achieve good, strong, consistent welds with good surface appearance.

The filler slug serves to keep the projection from collapsing and also supplies filler material to provide a good surface appearance. The filler slugs we used were simply scrap slugs from a separate punching operation. The method could be used with any aluminum that can normally be spot welded.

Robert Kayma, *Springfield, Ill*

11.25 'Strap' wrench won't mar parts

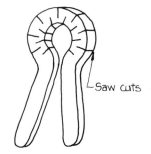

There comes a time when you have to grip a ground or polished diameter that can't be marred, the part has no wrenching flats, and there's no strap wrench available. I solved such a problem by making up a "strap" wrench from a piece of 1/2-in. aluminum plate.

With the aluminum plate clamped in the four-jaw chuck of a lathe, the hole was bored about 0.003-0.005 in. larger than the diameter to be gripped. Then a circle was scribed about 1 1/4 in. larger in diameter and two handles were scribed. This outline was then cut on a bandsaw. Slots 3/8 in. deep were then sawed alternately on the OD and the ID of the 5/8 in. wall thickness to give the wrench some flexibility.

Similar wrenches can be made to suit virtually any diameter.

JOHN URAM, *Cohoes, NY*

11.26 Tire plugs lock screws

We have some 18-station dial-type multi-operation machines in our plant, and we were having some problems with the nest positioning plates on the dials: vibration continually loosened the flat-head machine screws that held these plates down, thereby causing jamming problems during operation. We tried Loctite, dutch pinning, and bottoming the screws in the tapped holes, but these methods then caused difficulties in removal of the screws when the plates are periodically repositioned.

The ultimate solution was quite simple. We purchased 3/16-in. rubber plugs designed for plugging holes in tubeless tires, cut these to desired length, put them in the blind tapped holes, and tightened the machine screws down on top of them. The rubber plugs compress under the screws and absorb vibration to prevent loosening. Yet, when the plates have to be removed, it's easy to unscrew the fasteners.

Optimum length of the plug varies with the depth of the tapped hole, size of the screw, and composition of the rubber. However, we have found that for a 10-32 NF screw, the plug should be about 1/8 in. to 5/32 in. longer than the depth remaining between the end of the screw and the bottom of the tapped hole. A little experimentation should determine the proper length for other screw sizes.

WALTER ZBIKOWSKI, *Brentwood, NY*

11.27 Two ways to pull dowels from blind holes

Too often there's just no access from the rear for driving out a press-fit dowel. Here are two techniques that work.

The first method involves drilling and

reaming a 3/16-in. hole, although the diameter is certainly not critical, so it breaks through into the bottom of the dowel-pin hole. Fill the hole with oil to within about 1/4 in. from the surface. Then insert a 3/16-in. dowel into the hole and hit it with a hammer. It may take several fillings and hammer blows to completely remove a larger pin.

The second method involves the illustrated puller, which is made of tool steel, hardening optional. The center hole should be a sliding fit on whatever size dowel the puller is built for. It's set over the dowel against the part surface (with the jackscrews retracted), and the two clamp-screws are locked. Then the jackscrews are alternately tightened to pull the dowel.

DALE M GASH, *Blairsville, Ga*

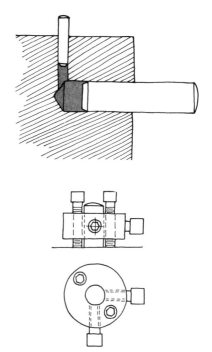

11.28 Weakened prongs ease assembly

We made a progressive die to produce a special 0.030-in.-thick washer with prongs on its ID to hold it on an aluminum shaft. During assembly, however, we found that the prongs were too stiff and that they scratched the anodized finish of the shaft. To cure this, we made a ring indentation at the base of the prongs to weaken them, which gained two benefits: assembly was made easier, and the anodized finish was left intact.

It would obviously have been simpler merely to stamp the washers from thinner stock, except that the 0.030-in. thickness was necessary for spacing.

ERNEST J GOULET, *Middletown, Conn*

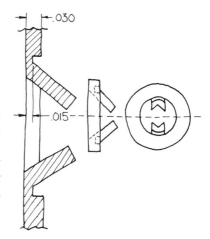

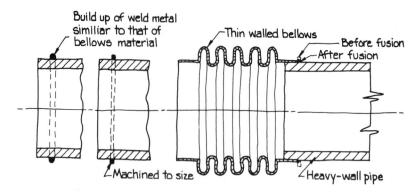

11.29 Welding thin-walled parts to heavy sections

If the thickness of the two pieces to be joined is about the same, welding of chromium- and nickel-base steels is relatively easy with TIG, MIG, or plasma arc welding processes. But if there's a considerable variation in wall thicknesses—such as the thin-wall bellows and the heavy-walled pipe shown in the sketch—then the job can be extremely demanding on the skills and experience of the welder. It's true that they could be silver-brazed, but welding is a must for the chemical-process lines in our case.

The problem, of course, is that when the current is set to suit the thin wall section, fusion between the mating parts would be improper. And when current was set for the heavy component, the thin one would quickly melt away.

The following technique was devised: First, weld material similar in composition to the bellows was deposited in a bead around the heavy pipe at the joint location. This was then machined so that the axial thickness of the ring was about the same as the thickness of the bellows, and the height was about two to three times the bellows thickness. The bellows was then slipped on the pipe, butted up against the machined bead, and fused together.

The method produces a leak-tight joint, and it eliminates the need for highly skilled operators. We've used the system successfully many times.

S RAM MURTHY, *Hyderabad, India*

12 Maintenance and Repair

12.01 Crane eases repair job on heavy tank

The problem was: how to handle a heavy, stainless-steel pressure vessel that had to be plasma-cut and welded in a repair job. The solution turned out to be both simple and highly satisfactory.

A heavy-duty dolly was inverted on the floor, and the tank was laid between the wheels of the dolly. Then a chain was wrapped around the tank and its hook was placed in one of the holes in the tank's flange. The free end of the chain was then connected to an overhead crane. Finally, the torch was set up at a convenient height on a mild-steel stand.

Switching on the torch while the rising crane slowly rotated the pressure vessel resulted in a nice, even, clean cut.

R K Das, *Hyderabad, India*

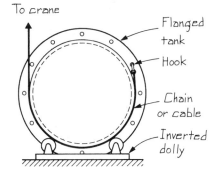

12.02 'Funnel' for small parts

All maintenance departments seem to keep their stocks of screws, washers, cotter pins, and other small components in a collection of cans, glass jars, or similar containers. The problem arises when it comes to returning the assortment to the containers after sorting out and finding just the one that was needed.

My solution was to make the simple

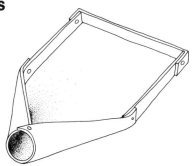

combination tray and funnel shown in the sketch. I used aluminum sheet, but a variety of other materials would be just as suitable.

D P VAUGHAN, *Newbury, Berks, UK*

12.03 How to re-zero dial caliper

Machinists' dial calipers frequently get some foreign particle in the rack that causes the indicator hand to jump teeth and give a false reading. Resetting the bezel to put zero somewhere other than "12 o'clock" is at best an irritating annoyance, so I use the following fix:

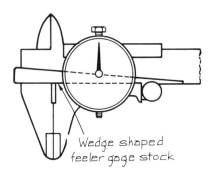

(1) Cut a wedge-shaped piece of feeler-gage stock or shim stock (0.008 in. to 0.010 in. thick) as shown in the accompanying drawing.

(2) Remove the bezel and crystal.

(3) Wipe the jaw faces clean and close the caliper.

(4) Insert the thin wedge behind the dial assembly and gently force the needle's pinion shaft against its spring load to disengage the pinion from the rack. Now return the hand to its normal 12-o'clock position and withdraw the wedge to allow the pinion to re-engage the rack. And finally replace the bezel and crystal.

H G ANDERSON, *St Paul, Minn*

12.04 'Replacing' O-rings on one-piece cylinders

Some air cylinders are made as a single unit, brazed solid, and are virtually impregnable against all efforts to replace worn O-rings. And a worn O-ring will leak compressed air, reducing its efficiency and causing the leaking air to deflect a die strip if the cylinder is being used as a misfeed detector.

However, you can defeat the impregnability by adding a supplemental O-ring instead of replacing the worn one. Just slip the new O-ring over the exposed piston rod and hold it in place with the simple clamp illustrated.

BEN SCHNEIDER, *West Orange, NJ*

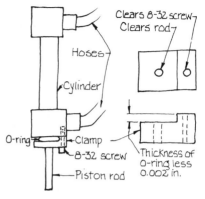

12.05 Split arbor pulls bearings in blind holes

Replacement of bearings mounted in housings with no access holes starts with the difficulty of pulling the old bearing—a task that is greatly simplified by this split arbor.

Using cold-rolled steel bar stock somewhat larger in diameter than the ID of the bearing to be pulled, turn one end to a slip fit (about 0.001–0.002 in. smaller than the bearing ID). Drill an axial clearance hole with a 60° countersink at the turned end. And split that end with a saw-cut about an inch long. Finally, turn a 60° angle under the head of a hex-head or socket-head bolt that's long enough to fit through the arbor. All specific dimensions depend on the job at hand—bearing size and accessibility, bar stock on hand, and the length of screws available. In some cases, a special screw may have to be made to fit the arbor.

To use the arbor, insert the screw through it and put a nut on the outer end. Then slip the arbor into the bearing and tighten the nut, which expands the arbor to grip the bearing's ID, and pull the bearing out.

JOHN URAM, *Cohoes, NY*

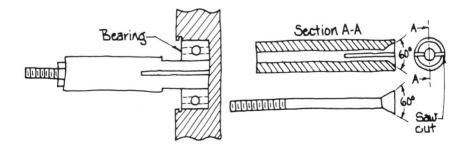

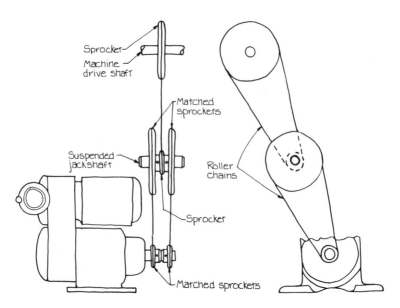

12.06 Suspended jackshaft for double reduction

In the process of rebuilding some equipment that was driven by a variable-speed motor drive unit through a roller-chain and sprockets, it was necessary to reduce the rpm considerably below what was possible with the slowest setting of the variable-speed unit. The problem was further complicated by three additional facts: (1) the smallest possible sprocket was already mounted on the drive unit; (2) the large sprocket on the driven shaft was enclosed, thus preventing installation of a larger sprocket there; and (3) there was no place to mount a conventional jackshaft.

The solution was to install a suspended jackshaft, as shown in the sketch.

A pair of matched sprockets is mounted on the drive unit to drive a larger matched pair on the suspended jackshaft that straddle a smaller sprocket that drives the machine shaft. All sprockets are keyed to their shafts, are properly aligned, and are spaced to provide adequate clearance for the central roller chain.

It would also be possible, of course, to construct a speed-increasing device with the same principle.

FRANCIS J GRADY, *Reading, Pa*

12.07 A valve indicator

Pipelines and plumbing seem to be proliferating in machine shops today to carry all

Maintenance and Repair

manner of fluids such as machining coolants, gas, and compressed air. We've found that we can save some time by adding position indicators to certain valves on these supply lines. They not only permit setting the valves to predetermined positions, but also allow a quick check from a distance.

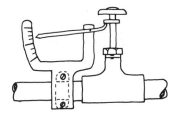

We merely add a collar below the handwheel and then clamp a sheetmetal unit to the pipe (see sketch). The pivot point of the lever is placed close to the valve stem to multiply the travel of the pointer end.

CLINT MCLAUGHLIN, *Jamaica, NY*

12.08 Worn lathe rebores its own tailstock

One of our larger lathes was beginning to give us trouble as a result of its worn tailstock. The quill was low at its front end; the rear was pretty good.

To correct this condition, we removed the quill and screw and made up a boring bar as shown. We fitted a cap, or bushing, to the rear end of the tailstock to provide support for the necessarily slender boring bar, which is gripped in a four-jaw chuck and indicated to bring it right on center.

The tailstock, which had been placed on the ways to the left of the carriage, is held by its regular clamp screw but with a spring under the nut so that it can be clamped firmly for machining, yet not too firmly so that it can still slide along the ways. Feed was provided by a bar clamped in the T-slot of the compound to push against the tailstock. The gearing was set for a slow feed of the carriage toward the headstock. We took two cuts through the bore to clean up the worn area.

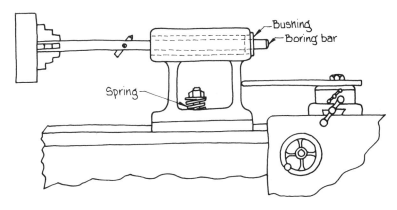

We then had the tailstock quill metalized, and then we turned it to a good fit in the new bore. The metallized surface is supposed to provide long wear, but even if it doesn't we can have it recoated. An alternate plan would be to bore out a larger ID in the tailstock and then press a suitable sleeve onto the quill.

CLINT MCLAUGHLIN, *Jamaica, NY*

13 Miscellaneous

13.01 Add a deburring station

It's very common in our tool and die shop to drill a few holes in a part and then ream them or tap them. That part is then ground and fitted, and then there are a few more holes to drill. Both ends of each hole have to be deburred, either by hand or with a countersink in the drill press—an extra operation that adds time.

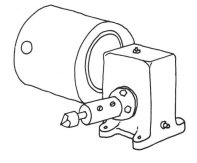

To speed this up, we mounted a small geared-head motor near the drill press and added a shaft coupling to hold a countersink, as shown in the sketch. For more speed and convenience, this is controlled by a foot switch. Because most of the pieces are small, it's very quick now the clean up the holes on this special deburring station.

CLINT MCLAUGHLIN, *Jamaica, NY*

13.02 Add another drawer

Space to store small tools is always at a premium. Here's a way to add another shallow drawer to a standard Kennedy toolbox.

The front cover of a Kennedy tool chest, when opened, normally slides under the bottom drawer for out-of-the-way storage. To utilize this space for additional tool storage, however, I made a drawer, as illustrated, out of 0.050-in. sheet steel that fits

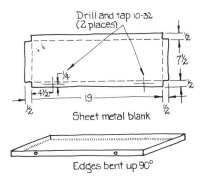

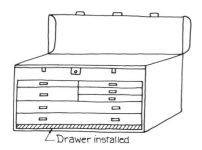
Drawer installed

in this space. Partitions can also be fabricated of the same material.

With the new drawer in place, the front cover can still be mounted so that the tool chest can be closed and locked. And when the toolbox is open, I just stand the cover behind it.

WALTER ZBIKOWSKI, *Brentwood, NY*

13.03 Bandsaw welder anneals spring ends

It's a simple process to make occasional springs by winding tempered wire on a mandrel in a lathe, which results in a typical plain-end spring such as shown at the top left on the drawing. But those springs often require special end configurations, either odd-shaped loops or hooks, or ground square ends like the compression spring shown at the top right. This often requires annealing the end coils.

The job of annealing can be done easily with the butt-welder commonly mounted on the column of a vertical bandsaw. Just make a pair of aluminum or brass plates 1/16 in. thick, 1/2-in. wide, and about 2 in. long and clamp them in the band-welder, one above and one below. Now set the selector switch to "anneal," hold the end of the spring against the plate as shown, and push the button. The end of the spring will compress axially to tighten the end coil, and it's a simple operation to grind the end flat manually.

BILL SAUER, *Hauppauge, NY*

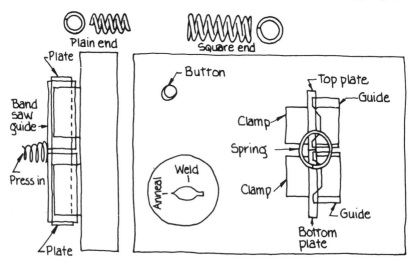

13.04 Block stores tools in sets

The tool-storage block shown in the sketch has saved a lot of time over the years by keeping the right size tap drill, clearance drill, and counterbore with the particular machine-screw taps that I most frequently use. The block can be made out of virtually any material, and the size depends on how many tap sizes are used most often.

WILLIAM HITCHEN, *Chicago, Ill*

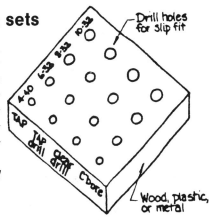

13.05 Brackets hold 'bedsheet' drawings

Working with long drawings or shop prints can be an annoyance. Commonly called "bedsheets," the unrolled ends of the drawing trail over the edges of the table and onto the floor—where they tend to get walked on, torn, creased, wrinkled, and dirty.

The simple rolled-drawing supports shown in the sketch eliminate all such problems. The rolled ends of the drawing are simply nested on both sides of the drafting table or shop bench. They're held below the table surface, where they don't interfere with the drafting machine or other equipment. And, when they're not in use, the supports can be swung away neatly under the table.

WILLIAM SLAMER, *Menomonee Falls, Wis*

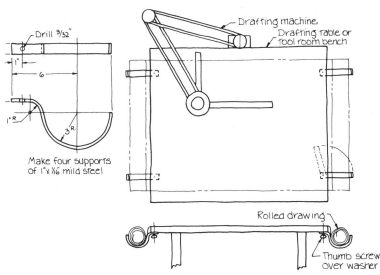

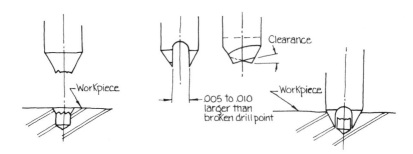

13.06 Broken centerdrill, reground, removes tip

It's never easy, and it often seems impossible, to remove the broken tip of a centerdrill from the workpiece. However, a fairly simple regrind of the broken shank, as shown in the sketch, can convert the otherwise useless piece of high speed-steel into a two-lipped trepanning tool that will core out around the broken tip. Because the same drill is being used, it will pick up the partially drilled hole and not wander.

RICHARD SCHULTZ, Tecumseh, Mich

13.07 Broken-stud removal

We recently had a problem with a 1-in. stud broken off about 1 in. below the surface. We made up a rough "bushing" by drilling a hole through a 1-in. length of 3/4-in. rod. This we dropped on top of the broken stud so it could be drilled accurately on center. Then we used an Easy-Out extractor to remove the stud without any difficulty. Total time was about 15 minutes.

CARL DeCOUDRES, Larned, Kan

13.08 Color-code your plant

We have taken your color-code idea one step further by color-coding both fixtures—pink for turret lathes, for example—and storage racks, helping us to monitor proper storage and to rapidly find "lost" fixtures.

E J ANNICK, Wellsville, NY

13.09 Coolant system gives balanced flow

The illustrated coolant system was designed to provide premixed coolant through existing water lines to individual coolant tanks. The design is such that no coolant can be isolated in one area, which would result in reduced coolant life. The balanced-flow system provides consistent pressure, so that

Miscellaneous

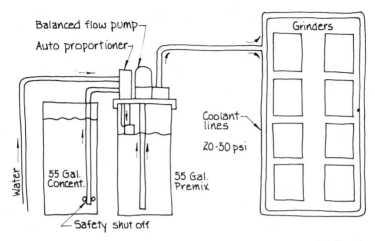

no ball check valves are necessary. As installed, it is capable of delivering up to 525 gpm, if required.

It works like this: When one or more coolant valves at the machines are opened, the pressure drops and the pump is energized to deliver the premixed coolant to the proper location (the open valve). When the 55-gal drum of premixed coolant falls below the required level, the automatic proportioner is energized, which recharges the premix drum to the proper level with the exact ratio of coolant and water.

This system has saved us a lot of production time, and it is an inexpensive semicentralized system to install.

BOB JOLLOFF, *Bluffton, Ind*

13.10 Crank mittens

Now that I've become accustomed to them, I doubt that I could still turn a machine-tool crank handle without a "miller mitten." This is nothing but a small, cloth bag that fits loosely, but without slack, over the handles on a milling machine's cranks.

A friend knitted a custom set for me, but a toddler's socks will do the job as well. The cloth has very little friction on the handle, making cranking much easier—surprisingly so.

DAVID R CARLSON, *Manchester, NH*

13.11 Ease that drill chuck off

Drillpress chucks can be ruined by hammering on them to get them off a tapered

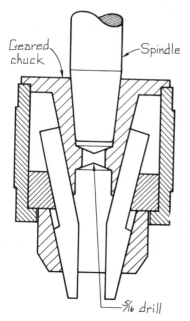

spindle. But drill chucks, such as the Jacobs chuck, can be simply modified for easy and safe removal.

Just drill a 5/16-in. hole axially through the web in the chuck body. Then a 1/4-in. hex nut can be chucked in the jaws, and a bolt screwed through this hole will contact the end of the tapered spindle and push it off the chuck.

A note on proper reinstallation of the chuck on the spindle: Clean the tapered surfaces, open the chuck to retract the jaws, and light application of a rawhide mallet will seat the chuck firmly on the spindle.

CHARLES M BARTLETT, *Birmingham, Ala*

13.12 Eyebolt fits three common threads

Too many times a heavy item of tooling—a lathe chuck, indexer, special fixture, or just a heavy vise—has to be moved somewhere in the shop or just be suspended from an overhead hoist for positioning and mounting and there's no suitable eyebolt to be found.

The sketch shows a quickly made solution to the problem that's even more utilitarian than a standard eyebolt because it provides three different threads. Popular thread sizes seem to be 3/8-16, 1/2-13, and 5/8-11. Check the equipment in your shop; you might have some odd ones.

It's a simple matter to cut a suitable length of heavy-wall tubing, drill it for the three capscrews, and then tack-weld them inside the ring so that they can be replaced easily if they are damaged.

DOUGLAS BAIRD, *Clinton, NY*

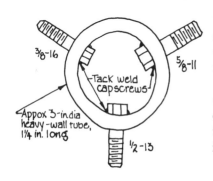

13.13 Flat belts from strapping tape

In our model shop it was necessary to run some test equipment by adding and subtracting various rotating components that were driven by flat belts. The problem was that we needed a large number of different belt lengths and we could not predict the required lengths until tests were partially complete. Then we would have to place a special order and wait several days for delivery. Furthermore, durability was important because life tests were part of the program; although loads were light, it was not uncommon to experience failure of the rubber belts before failure of the components under test.

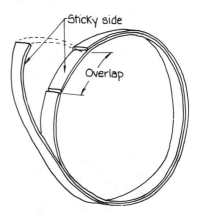

We solved the problems simply by making our own belts out of common strapping tape—the nylone-reinforced product used for binding packages. Two pieces of this tape were cut to the required length. Using care, we put the tapes together with the adhesive sides facing each other and making sure there was ample overlap. This gave us infinite flexibility as to the lengths we could use.

Although we initially considered this procedure an expedient to keep us going until an ordered belt arrived, we found that these belts survived the equipment under test. They are strong, yet light. The strapping-tape belts have proved to be so handy that we no longer order the rubber ones.

ART DRUMMOND, *Walworth, NY*

13.14 Heat-shrink part protectors

Heat-shrinkable tubing or sleeves, widely used in electrical and electronic work, can be used for protecting precision threads or ground surfaces that might be damaged in handling, shipping, or storage. The tubing comes in sizes from 3/64-in. ID to 4-in. ID, and is supplied in rolls from which any desired length can be cut.

The material shrinks to 50% of its initial diameter when subjected to temperatures in excess of about 250F. Recommended temperatures for maximum shrinkage are 200–300C (392–572F), and longitudinal shrinkage is less than 5%. To remove the tubing from a thread, it's easily screwed off.

JOHN R MAKI, *Beverly, Mass*

13.15 Holding parallels in a vise

When using thin parallels in a mill vise, they have an annoying tendency to fall over or move out of position. Lately I've been using a piece of steel strapping bent into a Z- or U-shape to hold the parallels in place with spring action. The method is fast, self-adjusting, and effective. Use 1/8-in. or 1/2-in. strapping for light jobs, 3/4-in. or more for larger jobs.

CLYDE COLLINS, *Seattle, Wash*

13.16 A 'household' convenience

Sometimes you need a small amount of lubricant on a bearing or cutting fluid at the tool tip. Ask your wife for an empty trigger-type plastic spray bottle (such as Glass-Plus, Windex, etc.) and fill it with oil. A little squirt will do the trick—safely—and without any need for a brush or oil can.

BEN SCHNEIDER, *West Orange, NJ*

13.17 Instant camera snaps setups

Many of the parts we produce in our shop require difficult and, often, original setups. And by the time the job comes up again, we have forgotten exactly how the setup was built. By the time we figure it out again, we've lost that much more valuable time.

Most machinists can't be bothered with drawing sketches or adding notes to the part print. So we purchased a Polaroid Instant Camera for the shop. Now all we do is take a picture of the setup with all the tools shown. We also put the job number at the bottom of the photo for future reference.

This has helped us more than we expected, because now the machinist encourages us to take pictures of "his setup."

RICHARD JORDAN, *Jordan & Smith Inc, Eastlake, Ohio*

13.18 Instant 'clamp'

Some parts just can't be clamped together with conventional holding devices for such operations as drilling through both. When such a job comes up in our shop, we use Super Glue (cyanoacrylate adhesive) to hold the parts together for machining. After the operation has been completed, the parts can be disassembled with a blow from a brass hammer or by prying them apart with a wedge.

DALE M GASH, *Otto, NC*

13.19 Job shop speeds quotes

Typical of most smaller job shops, we come across many varied jobs. When quoting a job that requires us to farm out OD or ID grinding, which we're not equipped for, it used to be necessary for us to estimate grinding costs ourselves (risky), to send a part print to the grinder (costly and time consuming), or to describe the part to the grinder over the phone (sometimes inaccurate).

To solve this problem, I made a chart with sketches of many different sizes, shapes, and configurations of typical parts we've done in the past. Copies were sent to each of our grinding sources, and we now phone the grinder, refer him to the proper sketch, and provide the necessary dimensions. This saves time and money in making quotations and ensures accuracy of communication.

RICHARD JORDAN, *Eastlake, Ohio*

13.20 Magnet prevents spills

Many toolroom operators have a small can of "special" cutting fluid that goes with them wherever they work, be the job on a lathe, mill, drill, or whatever. All too often, this can gets vibrated off its perch or elbowed over.

To help avoid making this kind of mess, I've found a magnet glued on the bottom of the can is a fairly effective way to stick it onto a machine table.

EDWARD LINDEMAN, *Rapid River, Mich*

13.21 Make a magnetic mirror

There are times when you have a job clamped to the table of a milling machine and it's necessary to do some operations at the back side of the workpiece. You have to stand in front of the machine to operate the feed and control handles, and you are unable to see the surface that you're machining. I remedied this problem with a discarded mirror from an old handbag of my wife's. Using an epoxy adhesive, I cemented a magnet to the back of the mirror. Now I can "stick" the mirror on the machine column and eyeball those previously invisible machining cuts.

JOHN URAM, *Cohoes, NY*

13.22 Masking-tape silencer

We have a horizontal milling machine that we use for machining plates square. The plates rarely come in flat, and the thinner ones—anything under 1 in.—generate a considerable amount of noise. I've discovered that the noise can be reduced to a quiet murmur simply by putting masking tape on the machine table.

WILLIE CROMEL, *Kenilworth, NJ*

13.23 Paper tray for teleprinter

Teleprinters are getting to be more common in manufacturing plants, and a typical method of feeding them their daily diet of paper is simply to put a box on the floor underneath. Too often this can get kicked out of place, causing misfeeds. Much neater, and quite simple to fabricate, it is a shelf installed just below the machine (ours is a Decwriter) with guide bars to align a sheetmetal tray for the paper. A slot in the front edge of the tray shows at a glance whether the paper supply needs replenishment or not.

JAMES SERRATORE SR, *Hatfield, Pa*

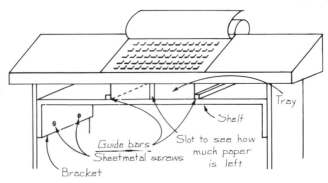

13.24 Peening straightens heat-warped parts

Parts to be surface-ground after heat-treatment are usually allowed some 0.015 in. excess stock for grinding. However, all heat-treated pieces do not come back from hardening as straight and flat as they were before hardening. Grinding such parts while warped will only result in undersize dimensions or hollow spots. Furthermore, such parts typically cannot be straightened by bending because of the danger of cracking the hardened metal.

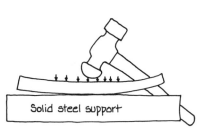

There is, however, a method for straightening such a piece. Place the part hollow side up on a flat block of steel and peen it all over the hollow area. This swages the material, forcing it straight, and the hammer tracks will be removed by the subsequent grinding step.

WILLIAM HITCHEN, *Chicago, Ill*

13.25 Plastic strapping makes handy shims

Black plastic strapping is often used today in place of the steel tape that was formerly almost universal for securing heavy crates, bundles of barstock, and even stacks of bricks. The plastic is tough, firm, and free for the taking and can be used for packing under clamps and chuck jaws. And the plastic surface won't mar even the most delicate work surface.

BOB MELVILLE, *Ithaca, NY*

13.26 Preserving soft hammers

There's a simple and inexpensive way to increase the useful life of lead, babbitt, brass, or any soft hammer made of other nonferrous metals. Whenever it's necessary to tap a drill, reamer, center-drill or other similar tool into a sleeve or collet, if the operator would merely tap a flat metal surface with the hammer a few times it would keep the faces of the hammer flat and serviceable for a longer time. In fact, we've noticed a 63% gain in the life of lead and other soft hammers in our shop.

MARTIN J MACKEY, *Parma, Ohio*

13.27 Prybar preserves chuck

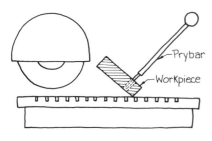

Simple things like this prybar can make your job a little easier—and at the same time help prevent scratching of the surface of your magnetic chuck.

The prybar I use is made of 3/8-in.-dia steel rod approximately 8 in. long with a 7/32-in. tip diameter and a black knob for a handle, although size and shape are entirely up to the individual and his typical workpieces.

Blocks of steel—especially hardened ones such as die parts—tend to stick to a magnetic chuck and to slide and scratch its surface when you try to remove them after grinding, even though the magnet has been reversed or turned off. Most of these blocks, however, have screw or dowel holes in them, into which you can insert the prybar so you can tilt it off the surface for easier and safer handling without scratching the chuck surface or the workpiece.

ANTON MESSERKLINGER, *Everett, Wash*

13.28 Quench it in steel

When thin blades of oil-hardening tool steel have to be heat-treated in your shop furnace and be held as flat as possible, there's a quick and easy way to do it. After reaching the required temperature, place the blades (one at a time) between two pieces of cold-rolled steel (as sketched) that are at room temperature and press downward lightly. The cool blocks of steel will absorb the heat of the workpiece in much the same way as an oil quench, and the blades will not only be hardened, but will also be flat and with minimum distortion. Then draw the temper as required.

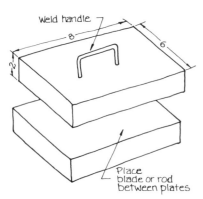

The method also works well with small

rods that have to be held as straight as possible, but, when applying the downward pressure, gently roll the rods back and forth until they're cool enough to touch. Result: hard (approximately 60–62 Rc) but straight rods. Again, draw as required.

ROBERT SANCHEZ, *Lockport, Ill*

13.29 Reflector aids visibility

When grinding draft clearance in a narrow slot in a die, it is sometimes virtually impossible to see how much "land" you're leaving. And you don't want to disturb the setup by lifting the die section off the magnetic chuck. Just slip a clean piece of white paper into the slot. It will act as a reflector for the light from the grinder's lamp, and will clearly illuminate the amount of stock you've removed. This trick works especially well if you're using layout blue on the surface being ground.

BEN SCHNEIDER, *West Orange, NJ*

13.30 Removing broken drill from a blind hole

When a drill breaks just below the surface in a blind hole or before it penetrates the far side of the work in a through-hole job, it's often possible to remove the broken tool with a length of welding rod. With the welding rod bent at an approximate right angle, its end is fused to the broken drill. The bent welding rod can now be used as a crank to rotate the broken drill point and ease it out of the hole.

MARTIN J MACKEY, *Parma, Ohio*

13.31 Retaining-ring hose-clamps

Small-diameter plastic tubing is put to many uses in the shop—air lines, vacuum lines, oil lines, etc. Often the tubing is simply slipped over the end of a smooth piece of copper tube, and after being in service a while the fit loosens. I've solved this problem by using TruArc retaining rings as hose-clamps. They do the job nicely, they take up very little space in tight places, and they're easy to apply and to remove later for reuse.

ANTHONY SIRACUSA, *Sewell, NJ*

13.32 Rough and ready tempering

For small pieces (under 3/4 in. long) of oil-hardening tool steel, you can shorten the time it takes for tempering after hardening in the oil quench. Merely lift the piece out of the oil and burn off the oil in the torch—but just barely burn the oil off—then quench it again. You'll have a piece tempered to Rc 59/61. The very first time you try this technique, practice with a small scrap of the oil-hardening steel. You'll find that *just* burning off the oil does the trick.

BEN SCHNEIDER, West Orange, NJ

13.33 Shortening screws on a lathe

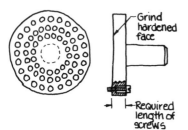

What machinist hasn't tightened a pair of nuts on an oversize screw, clamped it in a vise, hacksawed the screw to length, and then chamfered it manually on a bench grinder? The screws on hand just never seem to be the right length.

But our Assembly Dept recently needed several hundred No. 3-48 screws of a certain length—and the closest size commercially available was 3/64 in. too long. Special tooling for the shortening job was definitely in order.

We turned a 1/2-in. shank on a short length of 1 1/2-in. oil-hardening drill rod and then turned a flange as thick as the required length of the screws. Next we drilled and tapped a series of 3-48 holes in a pattern of concentric circles. To minimize the impacts of the interrupted cut, these holes were spaced as close together as possible, leaving only sufficient clearance for the heads of the screws. These holes were also counter-bored on the shank side of the flange to reduce the time needed for turning the screws in and out. And finally, after hardening the fixture, the outer side of the flange was skim ground so that the screws are deburred when they are removed.

The over-length screws are threaded into

the shank side of the flange and tightened. The loaded fixture, resembling a pin cushion, is then chucked in a lathe, and a facing cut is taken to shorten the entire batch of screws.

WILLY COLEMAN, *Whitestone, NY*

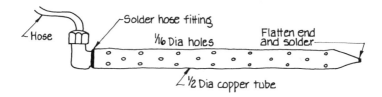

13.34 Simple aerator extends life of cutting fluids

I have found that a simple aerator, as shown in the sketch, will add considerable life to soluble-oil coolants in machines that may stand idle for weeks at a time. The one I made is connected to the shop air line through a small valve to control the flow of air, although an aquarium air pump will also do the job if shop air is not used. The aerator is then simply placed in the bottom of the machine's coolant sump. Results have been excellent in keeping the coolant from going rancid.

M PETER PASKA, *Lakeville, Mass*

13.35 Simple wire-straightener

You often want to use straight lengths of small-diameter wire that came in a coil. The photo shows a very simple straightener made out of about 6 in. of 1/8-in. brass tubing. The "kink" is offset about 3/8 in., and the right-hand (exit) end should be a very gentle curve.

To use it, chuck it in a lathe, feed the wire through the spindle, start the lathe at slow speed, and slowly pull the wire through the straightener with a pair of pliers. Result: straight wire.

MARTIN BERMAN, *Brooklyn, NY*

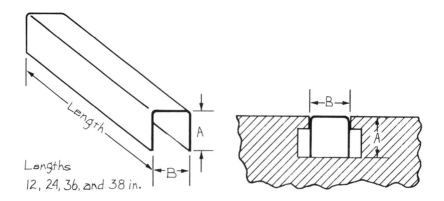

13.36 Slot covers speed cleanup

Easily fabricated T-slot covers will prevent a buildup of chips in the table T-slots of planer mills, horizontal boring mills, or similar machines, and thus expedite cleanup. Just bend them to fit the T-slots in your machine, using 22-gage (0.030-in.) cold-rolled low-carbon sheet or some similar material. It's convenient to make a variety of lengths to tailor them to each individual setup on the machine.

Peter Nahornuk, *Peabody, Mass*

13.37 Small syringe makes mini grease gun

I run a cylindrical grinder in a tool-and-die shop, and I'm frequently called upon to grind punches and die bushings with critical roundness, straightness, and concentricity tolerances. With these tolerances, there just isn't any room for dirt in center holes.

By using a small syringe as a miniature grease gun loaded with center lubricant, I'm sure that there's never any dirt in the grease. Additionally, the syringe makes it very easy to put the grease exactly where it's needed and in exactly the amount I want. Not only don't I have to worry about problems arising in the middle of finishing operations, the grease even lasts longer.

Thomas R Boxler Jr, *Waynesboro, Va*

13.38 Soft tube is file cleaner

The age-old problem of file cleaning encountered by every machinist can be remedied by a simple, inexpensive tool made easily from a piece of 3/4-in. copper water tubing. Just crimp about 1/2 in. of one end flat in a vice and file a 30° bevel on it. While you have the file in your hand, also carefully deburr the other end for safety.

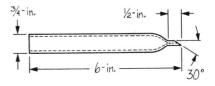

By slowly drawing the tool over a clogged file in line with the teeth, all debris is removed—leaving a very clean and unobstructed tool. Periodic resharpening of the tool will be necessary because of the softness of the copper and the cleaner's use on different styles and grades of files.

JERRY RUSSELL, *Troy, Ohio*

13.39 Soldered rod reinforces sheetmetal bend

An L-shaped angle baffle made of thin sheetmetal is often used to deflect parts as they come off forming tools or to break the fall of parts down a chute on the way to a catch box. Especially in long-run production, these bent baffles will sometimes crack at the bend line as a result of metal fatigue.

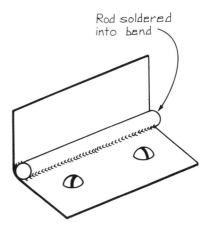

To avoid this problem, I reinforce that bend line by soldering a length of soft steel, about 3/32 in. diameter, into the inside corner of the L. An 8-penny nail is fine for the job, as it accepts solder easily, if the head is cut off to eliminate any interference.

BEN SCHNEIDER, *West Orange, NY*

13.40 Speedy table cleanup

Considerable time can be spent in cleaning out the table T-slots on milling machines, jig borers, and other machine tools. To speed this chore, we recently cut a piece of sheet metal to the same size as the table (or larger if you want to fold the edges down) and drilled two holes in it to align with the hold-down bolts of the vise.

We first slipped the two bolts into position in the T-slots, then placed the sheet-metal table-cover over them, and finally clamped down the vise.

Now when a job is finished it only takes a couple of wipes to clear the chips off the table instead of tedious work to clear them out of the T-slots.

DONALD E VAN HUIS, *Holland, Mich*

13.41 Spool guides tool

An ordinary wooden spool from your wife's sewing box makes a handy guide to hold a drill perpendicular to a surface you are drilling with a hand drill—and there are lots of times when you just can't get the workpiece onto the table of a drillpress. Alternately, you can take a flat block of wood, drill it accurately on the drillpress, and then take that piece to the work for drilling a perpendicular hole.

The technique is also useful for starting hand taps true to the hole.

JOHN R MAKI, *Beverly, Mass*

13.42 Sticky tape removes filings from magnet

Removing iron filings or chips from a magnet can be a frustrating chore—but it can easily be accomplished with Scotch or masking tape.

W E TRITZ, *Waukesha, Wis*

13.43 Straightening warped hss

Blade(s)

We use hardened, thin (0.045-in.) blades of T15 high-speed steel in a slot-broaching operation. A quantity of blades that were on hand were bowed out of flat and could not be finish-ground to the precision dimensions required. Attempts at cold-straightening were unsuccessful—because of their high hardness (68 Rc) they sprung back to the original bowed shape every time.

Key to straightening the blades was to heat them under load in the fixture shown at 1000F for about 30 minutes. This temperature is okay for hss, and the loss of hardness is negligible. With the correct deflection of the blades in the fixture, flat parts were obtained consistently.

ROBERT J MCMASTER, *Winchester, Mass*

13.44 Tap generates worm-wheel

Small worm-wheels can be cut rapidly on a lathe using a standard tap as a hob. The worm-wheel is then used in conjunction with a standard machine screw for the worm to make a compact mechanism for very fine movements.

The tap is held in a lathe collet and is rotated at a surface speed of about 6–7 sfm. The worm-wheel blank is held on a freely rotating vertical spindle in a fixture block mounted on the cross-slide. The blank is fed rapidly into the tap by an amount slightly less than the depth of the thread on the tap, and the tap, of course, then drives the blank to cut successive teeth. It takes about two revolutions of the worm-wheel to finish the teeth. The method has been used successfully to cut worm-wheels of brass, bronze, and soft steels. An ample supply of cutting oil will improve the finish.

To calculate the blank diameter, multiply the number of teeth on the worm-wheel times the tap pitch, divide this by 3.1416, and add the depth of the thread.

G Sudheendra, *Bangalore, India*

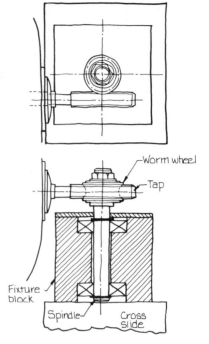

13.45 This key has knobs

I find that knobs on the crossbar handles of drill chuck keys make gripping them much more comfortable and secure, especially for tightening tools in lathe tailstocks. The knobs are retained by setscrews.

John Speer, *Stamford, Conn*

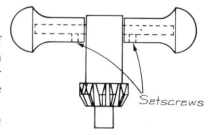

13.46 Toothpaste for lapping

We recently had an unusual job requiring a metallic seal between a soft, low-carbon steel part and a 304-stainless component. Specifications required that the seal be tested at 4260 psi hydraulic pressure.

Lapping the seal with common emery compound left minute grooves in the seal surfaces, and it leaked at test pressure. Finally, we lapped the seal with ordinary toothpaste, and it not only handled the required test pressure, but actually withstood 4970 psi for 24 hr.

C R NANDA, *Bombay, India*

13.47 When the collet won't fit

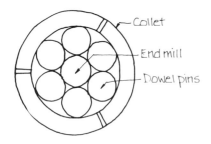

We had a project that required some machining with miniature end-mills (3/16-in. shank) on a collet-equipped machine for which no small collets were readily available. Adapters or bushings seldom run true enough for very small end mills, but we improvised an effective solution that ran as true as the collet we had, and required only some standard dowel pins.

For the 3/16-in. shank, we used six 3/16-in. dowel pins of appropriate length spaced radially in a 9/16-in. collet, carefully avoiding positioning any of them at the collet slits. This leaves a center space that accommodates a 3/16-in. diameter. In most cases, the dowels will fit snugly and won't fall out even without the end-mill inserted.

The same method works with 1/8-in.-shank tools by using 1/8-in. dowels in a 3/8-in. collet; and for 1/4-in. shanks, 1/4-in. dowels in a 3/4-in. collet will do the job.

FREDERICK C LORENZEN, *Lorenzen's Tools & Dies, Bellmore, NY*

13.48 When you need a thin shim

When you need a really thin shim, for example to put on one side of an end-mill shank to produce a tiny bit of runout for milling a keyway a few "tenths" oversize, use a bit of cellophane such as that from a cigaret package.

SCOTTIE COLLIGEN, *Augusta, Ga*

13.49 Window shades roll blueprints

Ordinary window shades make excellent props for blueprints and part drawings. The shades come in various sizes, they're inexpensive, and they take up little space. Drawings can be fastened to them with masking tape and can be rolled up when not in use. The shades can be mounted to a wall with brackets, suspended from low ceilings with light suspension arms, or they can be mounted to machinery with suitable brackets.

WILLIAM SLAMER, Menomonee Falls, Wis

13.50 Wooden 'soft jaws' aid many vise jobs

Cutting off the end of a small, short pin or fastener—as one example—can be made easier and safer by clamping the part in a bench vise along with two pieces of soft wood, as shown in the sketch. It firmly secures parts of irregular shape, it provides an easier start for the hacksaw blade, and it even holds the cutoff end after it's been cut.

JACOB SCHULZINGER, Houston, Texas

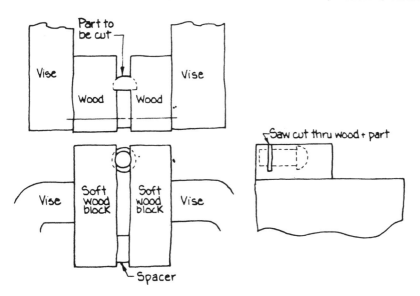